More Than Meets the Eye

More Than Meets the Eye

AN INTRODUCTION TO MEDIA STUDIES

GRAEME BURTON

A member of the Hodder Headline Group
LONDON • NEW YORK • SYDNEY • AUCKLAND

First published in Great Britain in 1990
Second edition published 1997 by
Arnold, a member of the Hodder Headline Group
338 Euston Road, London NW1 3BH
175 Fifth Avenue, New York, NY 10010

Distributed exclusively in the USA by
St Martin's Press Inc.,
175 Fifth Avenue,
New York, NY 10010

British Library Cataloguing in Publication Data
A catalogue entry for this book is available from the British Library

Library of Congress Cataloging-in-Publication Data
Burton, Graeme.
More than meets the eye : an introduction to media studies / Graeme Burton.
p. cm.
Includes bibliographical references and index.
ISBN 0–340–67663–9
1. Mass media—Study and teaching. I. Title.
P91.3.B79 1997
302.23'07—dc21 97–961
CIP

ISBN 0 340 67663 9

Composition by Phoenix Photosetting, Chatham, Kent
Printed and bound in Great Britain by J W Arrowsmith Ltd, Bristol

Contents

Acknowledgements

I would like to thank colleagues and students present and past who have helped make this book, even though they don't realize it. In particular, I would like to thank Richard Dimbleby for our times practising together this writing business, and my sympathetic editors for their patience with my delays and for having a sharp eye for error. And a personal thank you to Judy for being there when the spirit was weak.

The author and publishers would like to thank the following for permission to use copyright material in this book: the Advertising Association for figure 6.1; Aston Business School for figure 2.1; © 1994 BBC Education for figure 5.11 (Middlemarch); the Broadcasters Audience Research Board Ltd for figure 8.9, and the Broadcasters Audience Research Board Ltd and Taylor Nelson AGB plc for figure 9.2 (*Spectrum*, Spring 1995); Continental Research for figure 4.15 (1995 Internet Report); the Department of Health and Social Services for figure 5.6; Egmont Fleetway Ltd, 1997, for figure 6; Film Education for figure 4.3, adapted from a Film Education pack, © Film Education and Diet Coke, and figure 8.2, adapted from a Cinema 100 Education Programme booklet; *The Guardian* for figures 4.2 and 5.8; HarperCollins Publishers Ltd for the front cover of *The Mirror Crack'd from Side to Side* by Agatha Christie (figure 5.2b); Hutchinson for the front cover of *Going Wrong* by Ruth Rendell (figure 5.2c); *The Independent* for figures 4.10 and 9.1; the Independent Television Commission for figures 1.2, 1.3, 4.14, 5.3 and 8.6; Manchester University Press for the tables and pie charts from *Mass Communication* by Rowland Lorimer (figures 4.11, 4.12 and 4.13); Marvel Entertainment Group Inc. for figure 5.13; models Sarah Maltravers & Jason Bailey, MOT Model Agency, and the photographer John Wallace, for figure 2.2; Mysterious Press UK for the front covers of *The Neon Rain* by James Lee Burke (figure 5.2a) and *The Big Nowhere* by James Ellroy (figure 5.2d); the Office for National Statistics for figures 8.7, 8.8, 8.10, 8.11 and 8.12 (*Social Trends 26*, © Crown copyright, 1996); Routledge/Methuen and Co for figure 1.1, adapted from *Power Without Responsibility* by J. Curran and J. Seaton (1980); *Sugar* magazine, AtticFutura Publishers, for figure 8.5; and Terry Williams for the mischievous drawings.

Every effort has been made to trace all copyright holders. The publishers will be glad to make suitable arrangements with any copyright holders whom it has not been possible to contact. Any rights not acknowledged here will be acknowledged in subsequent printings if notice is given to the publisher.

Introduction

TO THE READER

This book is for students of the media following a variety of courses. It is designed to help make sense of the media, to explain how the media communicate with us, and to give you an understanding of how they operate and of who runs them.

I assume that you like reading, watching and listening, and that you are curious about what goes on in the media world, because it is a part of all our worlds, of all our experience. The following chapters should cause you to think about the media and media products in new ways.

THE AIMS OF THIS BOOK

I want to help you make sense of what is going on in the media, and to add to your interest in and enjoyment of what you watch, listen to and read. The following chapters should help you organize your own ideas about the programmes you watch, about the newspapers you read and about all the media which are part of our everyday lives. You should come to understand that there are reasons why we have the kind of media that we do have in Britain, and why it is important to study them.

This book will explain how media organizations put their messages together, why they do this in the way they do, and what effects their output may have on us.

In particular I want to clarify some of the terms which are commonly used to make these explanations. Indeed, I want to explain why these terms and ideas matter anyway. So I hope that you will understand more about the media in relation to your life, your thinking and your world by the time you have finished this book.

Most of all, as a student of Media and of Communication, you should be encouraged not to take things for granted. Books, records, programmes do not just appear from nowhere for no reason. The reasons why they appear in the way that they do are important, because they help shape our view of the world. This book aims to help you understand this process of shaping.

THE STRUCTURE OF THIS BOOK

The book is organized into nine chapters. The first three deal with basic ideas about the study of media in general. The rest follow through media communication from where it starts – with the owners of systems and makers of programmes – through to where it ends, in the heads of ourselves, the audience.

Chapter 1 asks the basic question, why bother to study the media at all? It gives you some pointers to what you may get out of media study besides straightforward enjoyment.

Chapter 2 gives you some particular ideas about how you can study, how you can analyse media material and avoid the trap of just chatting about papers and programmes in a rather general way.

Chapter 3 presents the media as examples of communication like any other – this book, for instance, or even talking. So it goes into the idea of communication as a process which has various parts. The remaining chapters will help you to understand the parts of this process.

Chapter 4 deals with the organizations that make the media material. They decide to pay for and to manufacture media material in the first place, so we need to see why and how they do this.

Chapter 5 is about the way that this material is made and about the sort of material that gets made. It asks why so much media product is basically the same, why we get so many stereotypes in this material, and how this affects us. Can we say that some media products are more or less realistic than others?

Chapter 6 is an extension of 5, looking at News and Advertising in particular. It explores how we are led to understand news material in one way rather than another. It talks about some basic aspects of advertising and about devices of persuasion which appear in adverts.

Chapter 7 looks at the messages in the programmes and magazines, and other examples of product. Something is being communicated to us, so what is it? Even casual entertainment, even the most fun shows are telling you something, whether you realize it or not. And what you are being told really matters because it goes inside your head.

Chapter 8 focuses on the audience. We need to look at ourselves as consumers of the media products. Many of the media makers certainly regard us in this way. So how exactly do they see us? How does this affect what they put out?

Chapter 9 picks up on Chapter 6 in terms of the possible effects of all this material upon us, on the way that we think.

So this is what you are in for if you read on. The book is not just a lot of facts about the media but more about ways of using those facts, about ways of making sense of the media.

HOW TO USE THIS BOOK

Basically, you should use it to find out what you want.

I have written the book in certain chapters in a certain order because I think that it makes sense. But it is your book. So you should use it the way that you need to – the way that you should use any book.

If you aren't bothered about reasons for taking a media course then skip Chapter 1. If you don't want to be held up by the introduction to study methods, then leave these until the end. If the explanation of the basis for Media

Studies as a communication process looks too heavy, then leave that for later. It won't matter, though I hope that you will read it all at some time.

Most of all, you should learn to pick out bits to suit your interests and study needs. If you are on a taught course then your teacher will have organized a course plan, with topics in a given order. Look at the chapter details. Look at the table of contents. Look up the key words for the topics that you are dealing with. Go to the parts that you need.

Do use the reading list, which might help you chase up ideas and topics.

Do use the glossary, which will help you sort out the meaning of special words and phrases that are used.

By the end of it all you really should have a better idea of what is going on behind the act of watching a programme or reading a newspaper. For factual details about how a newspaper is made or how special effects on television are done there are other books – and your teacher – that can deal with such matters. This book focuses on your ideas and understanding. It will enable you to study how and why the media create the things that they do.

DAILY MAIL
VENUE
COSMOPOLITAN
ECONOMIST
FILM
GUARDIAN
BEANO
THE TIMES

1

Why Study the Media?

There should be reasons why you study anything. These reasons should be ones that you understand and believe in. You should know why you are doing anything, in fact. And those reasons should answer that big question – SO WHAT? This is Media Studies – so what?

So think about the following answers to that question. I hope they will convince you, and that some of them will be things that you agree with.

1 YOUR INTERESTS

In the first place you may be personally interested in certain types of magazine or programme, so you want to look into these further. You may also be interested in how the material is created; you may like reading about the mechanics of media making. Again, you may feel that you want to study media because it is something you know about, something you feel confident with, a subject where you have a head start because you already have some background in it and some opinions.

2 MEDIA POWER AND INFLUENCE

Many of the main arguments for studying the media come down to this: everyone believes that they do have some power, though it is surprisingly difficult to establish exactly what kind of power this is. The main power of the media lies in the fact that they can shape what we know about the world and can be a main source of ideas and opinions. They may influence the way we think and act. This power is the greater if we take the media together rather than looking at one individual medium such as television. And it is most obvious when we look at examples of media use such as an advertising campaign. Such campaigns do use media together, and thus repeat and reinforce any message they are putting across. Some people argue about how much power the media really have. But the continuous and public arguments suggest that it must be worthwhile studying the media in order to see whether or not they really do have this power and influence.

3 ECONOMIC POWER

In particular, the sheer economic power of the media makes them significant and worth studying. The media industries employ thousands of people directly and make the employment of thousands of others possible (for example in terms of production of equipment). The income and expenditure of the media are vast. The amount spent on making programmes or producing a magazine is colossal. For instance, the total income of the ITV companies was £2 billion in 1994, of which about three quarters was earned by Channel 3. The licence fee income of the BBC was £1,819,000,000 (or about one and two-thirds billion pounds) in the financial year 1995/96, after deducting the cost of collecting it. The record industry sold nearly one and one-third billion pounds worth of discs and tapes in 1995.

And then what about advertising? In 1995 Britain spent £8 billion on advertising – which supports the media. People do not spend this sort of money on making messages and sending communication unless they believe that it has

Percentage of newspaper circulation figures held by the four leading corporations

Corporations	Percentages
Mirror Group Newspapers *The Mirror, Sunday Mirror, People*	National daily papers 84%
United Newspapers *The Express, Daily Star, Sunday Express*	
News International *The Sun, The Times, The Sunday Times, News of the World*	National Sunday papers 88%
Associated Newspapers *Daily Mail & Mail on Sunday*	

Market share of leading five companies in various media

Channel 3 companies (new programmes × hours)	77%
National daily newspapers (circulation)	91%
National Sunday newspapers (circulation)	91%
Books (sales × millions of pounds)	82%
CDs, cassettes (sales × units)	60%
Video cassettes (× units rented)	66%

Fig 1.1 Media ownership – dominant market shares

some effect. So the sheer financial clout of the media is an argument for studying how and why they work the way that they do.

4 SCALE OF OPERATIONS

This is also huge. It provides an argument for looking at what is going on. One may measure this scale in all sorts of ways. One could talk about it in terms of size of audience. Peak time audiences for television run to about 18 million people – that's over a third of the population of this country. One could look at it in terms of geographical scale of operations. So what about satellite channels like Sky, spraying down their programmes over Britain and most of Europe? Or what about scale in terms of proportion of the particular media industry controlled? Nintendo and Sega have the cartridge computer games industry nearly sewn up. Nor is it just about production. If you want to launch a new magazine on a national scale in Britain, then you need to be sure of the retail support of WH Smith, with its dominance of prime site outlets, let alone its control of nearly 50 per cent of distribution business. However we measure this scale, this massness of the mass media, it is so great that again it must make the media a subject worth looking into.

5 ACCESS TO THE AUDIENCE

This is about the ability of the media to get to their audience. It is notorious that radio and television are special in the sense that they get into the living room. They have access to every household in the land. But then again there is *The Sun*, which reaches nearly 4 million people every day. Or *The News of the World* which gets to 5 million people every Sunday (and both papers are owned by the News Corporation). This ability to get to us, sometimes in huge numbers, also makes the media special as communicators.

6 INFORMATION AND ENTERTAINMENT

These two things are very important to most people. So where do we get most of this from? Right – the media. Every week, about 650 hours of television are put out (excluding satellite transmissions). Every year, some 220,000 hours of television are pumped out through all channels. There are 75 licensed satellite channels putting out about 200,000 hours per year. Ninety per cent of people polled say that they get their news from television, and that they believe it for the most part. Thirteen million people read the three main tabloid and three main quality newspapers every day. There were 130 million cinema attendances in 1995. If you think about it, where else do you get your information and entertainment from? Remember, books are as much a media industry as any other. You may spend time socializing and playing a sport. But again, statistically you are spending an average of two hours per day just watching television – that's 750 hours per year. So if the media are main providers of information and entertainment, again it makes sense to look at what they provide and why, because it certainly is a major part of your life.

7 REPETITION OF MESSAGES

This refers to the repeating of items of information and entertainment. The main channels of serious news in three of the media all give you more or less the same items, much of it from three main news agencies. The fact that thousands of copies of newspapers are printed means that the same messages are being reproduced thousands of times. Films appear in the cinema, a year or so later they come out on video, about two years later they appear on broadcast television. Now they are appearing on satellite and cable systems.

	1988 %	1989 %	1990 %	1991 %	1992 %	1993 %	1994 %	1995 %
Video recorder	58	70	75	72	77	80	83	84
Teletext	25	30	41	40	46	49	55	60
Compact disc	n/a	n/a	n/a	22	29	27	35	37
Home computer	18	26	22	22	30	29	29	27
Video games	10	11	11	12	17	20	23	23
Satellite TV dish	*	2	4	8	10	12	14	15
Video camera	2	3	4	5	9	10	11	14
Nicam digital stereo television set	n/a	n/a	n/a	5	6	7	10	10
Cable TV	1	2	1	3	3	3	7	6
Widescreen television	n/a	n/a	n/a	n/a	n/a	n/a	7	6
Cable phone	n/a	n/a	n/a	n/a	n/a	1	5	6
Video disc player	1	2	2	2	3	2	4	5
Have one or more of these	67	77	81	80	83	86	89	91
Have none of these	33	23	19	20	17	14	11	9

*Base: All TV viewers. Note: 1. *Less than 0.5%. 2. n/a Not asked*

Fig. 1.2 Home entertainment equipment in the home (all viewers)

Figure 1.2 shows how important and how available technology is now, in terms of making mass media immediately accessible to large audiences.

Whatever one gets out of a particular film, this too is being repeated across various media. And there is evidence to prove that what is repeated is believed. So this, too, is a good argument for studying the media and for studying the possible effects of this repetition.

8 MEANINGS AND MESSAGES

All this media material tells us something. On one level there is often a pretty obvious intention to tell us something – perhaps tell us what a new record release is like through a review. But on another level what we are told is less obvious. It is, perhaps, not so obvious that an advertisement warning against AIDS is actually suggesting that women are significant AIDS carriers (because one particular campaign featured a photograph of a young woman). This

	Often %	**Occasionally** %	**Rarely** %	**Never** %
I watch the same programmes because I like them and know when they are on	72	21	5	2
I make selections from *TVTimes, Radio Times* or some other TV listings magazine	49	22	16	12
I read the TV listings that appear in the newspaper	47	28	14	10
I watch programmes picked by other family members or people in my house	25	40	18	16
I skip from channel to channel until I find something interesting	23	36	27	13
I choose on the basis of advance programme trailers shown on TV	19	52	22	7
I plan what to watch several days in advance	16	27	32	25
I follow recommendations of friends	15	46	28	10
I watch one programme and then leave the set tuned to the same channel	12	31	35	21

Base: All TV viewers. Note: 'Don't knows' have been excluded

Fig. 1.3 Why people watch what they do

Figure 1.3 illustrates the importance of the 'why' question in any course of study. Why do we watch what we watch? Why does it matter? Why do some people have some ways of selecting media material, but not others? Item one in the table begs the huge question of why people like some programmes and not others. Study of the media involves an attempt to answer such questions.

example is important because it is simply untrue that females are significant in this way. Statistically it is males who are more likely to be carrying the virus. So a good argument for studying the media is to dig down for these less obvious meanings and messages, and then think why they are there anyway.

There may be other reasons why one should study the media. But those just given provide some major answers to the title question of this chapter. They provide a good argument for looking into what is being said, who is saying it, how it is being said, who is taking in what is being said, and what effect all this may have on readers and viewers. So I hope that you feel convinced, if you needed convincing, that what you are doing is worthwhile and important to you as an individual.

REVIEW

You should have learned the following things from this chapter.

That there are a number of good reasons for studying the media, which include

- personal interest
- belief in the power and influence of the media
- the facts that the media have great economic power, that they are our main source of information and entertainment and that they construct many messages which may influence our views of the world.

Activity (1)

This activity is concerned with the scale of operation of the media, with the influence media may have, with the production of meanings by the media.

It focuses on computer games, and on the dominant position of Nintendo and Sega in particular.

You need to acquire or have access to magazines about computer games, and to identify at least 10 leading games, knowing something about what is involved in them.

Examine the general content and treatment of the games and the magazines, and ANSWER THE FOLLOWING QUESTIONS –

- who owns the games, including those referred to in the magazines?
- what are typical sales figures for leading computer games?
- what is the subject matter/'storyline' for these games?
- are there similarities between the games in these two respects?
- what are the roles played by women?
- what are the roles played by characters who are not white in terms of ethnic origins?
- which games are themselves spin-offs of other examples of media product?
- which games have other connections with other examples of media products?

The answers to these questions should prove a point about the power of big media corporations, about the scale of their production, about the links between media industries, about the ways in which repeated meanings about things like race and gender are put across by media products.

MAKE A COMPARISON between platform games and interactive games such as *Monkey Island* and *Simcity*.
Your comparison should concentrate on how the game player is involved and can use the game, and on differences of general treatment in each case.
You could set out your points as a list of ideas, to the left and right of a line.

CONDUCT AN INFORMAL SURVEY of females, asking them why they don't like or play such computer games.
Discuss your findings afterwards.

TRY DESIGNING A GAME FOR GIRLS
Just describe the main aspects of theme, character if appropriate, and what the game player does.

Reflection on these activities should throw up some ideas about the relationship between media and audience. Also you will learn something from discussing why it is difficult to design a computer game that will attract females. This also leads to ideas about ideology, about views of the world which exist within media and which are communicated to us through the media.

BBC 1

6.0 **BUSINESS BREAKFAST** (94446).
7.0 **BREAKFAST NEWS (T)** (25717).
9.0 **BREAKFAST NEWS EXTRA (T)** (2134953).
9.20 **DELIA SMITH'S SUMMER COLLECTION (T)** (rpt.) (S) (7751224).
9.50 **GOURMET IRELAND:** With chef Aidan McGrath (S) (5195224).
10.20 **THE FLYING GOURMET'S GUIDE TO THE GREAT BRITISH BIRD TABLE** (7375576).
10.50 **NEWS; REGIONAL NEWS AND WEATHER (T)** (5431798).
10.55 **CRICKET: SECOND TEST:** England v Pakistan. Live coverage. Continues on BBC2 (S) (5395137).
12.0 **NEWS; REGIONAL NEWS AND WEATHER (T)** (1887040).
12.5 **THE ALPHABET GAME:** Fun-filled word game (S) (4057243).
12.35 **NEIGHBOURS (T):** Phil worries Helen intends to hang up her paint brushes for good, while Brett decides to give Judy another chance. Stonefish attends an acting audition (S) (6895446).
1.0 **NEWS (T); WEATHER** (95576).
1.30 **REGIONAL NEWS AND WEATHER** (45162934).
1.35 **CRICKET: SECOND TEST:** England v Pakistan. Further ball-by-ball coverage of day one from Headingley. Continues on BBC2 (S) (73244717).
5.35 **NEIGHBOURS (T)** (rpt.) (S) (538972).
6.0 **NEWS (T); WEATHER** (427).
6.30 **NEWS WEST** (779).
7.0 **HOLIDAYS OUT (T):** Report from the World Rowing Championships at Strathclyde Country Park, featuring a visit to a cowboy ranch in Ireland and a day trip to Boulogne (S) (4595).
7.30 **EASTENDERS (T):** David tries to convince Lorraine that Joe's welfare is now his first priority, while troubled couple Carol and Alan clash with rebellious daughter Sonia (S) (663).
8.0 **● BACK TO THE WILD (T):** See Pick of the Day (S) (3243).
8.30 **AUNTIE'S SPORTING BLOOMERS (T):** Another collection of sporting gaffes and comic mishaps both on and off the pitch, with Terry Wogan (S) (2750).
9.0 **NEWS (T); REGIONAL NEWS AND WEATHER** (3088).
9.30 **ATLETICO PARTICK (T):** Jack is relieved to be dropped from the team, as it means he doesn't have to face Mambo's dirty tactics — and his luck improves still further with the arrival of a new female physio. With Gordon Kennedy (S) (93021).
10.0 **● DEFENCE OF THE REALM (T):** New series. Cameras go behind the scenes at the Ministry of Defence, filming the daily working lives of defence chiefs and ministers. See Best Documentary (S) (640601).
10.55 **FILM (T): Crazy People (1990):** When an advertising executive starts writing ads that tell the truth he is sent to a sanatorium by his partner. Screwball comedy, starring Dudley Moore and Daryl Hannah (S) (4666408).
12.25 **FILM*: Invasion Of The Saucer Men (1957):** Newly-weds accidentally run over an alien and unleash a reign of terror. Sci-fi comedy, starring Steve Terrell and Gloria Castillo (3633625).
1.30-1.35 **WEATHER** (6015199).

PICK OF THE DAY

● Back To The Wild, BBC1, 8pm.
Hoping to repeat the runaway success of *Animal Hospital*, which centred on the poorly pets of North London, this series looks at another RSPCA hospital, West Hatch in Somerset. Here, the patients are all wild and it is the job of the vets, nurses and helpers to get the animals fit enough to return them to their natural environments. Prepare to have your heartstrings tugged as the vets struggle to keep alive an orphaned roe deer fawn, baby badgers, fox cubs and ducklings. Patrick Robinson (Ash in *Casualty*) introduces the stories. Whether or not this five-part series will emulate *Animal Hospital*'s success is questionable. Robinson is no Rolf Harris; he's merely there to bookend the episodes. Nor does it have the spontaneity of the live series — but then sick, furry animals are always a ratings winner. FRANCES LASS

BBC 2

7.15 **SEE HEAR BREAKFAST NEWS (T)** (6241798). 7.30 The Brollys (1610363). 7.45 Lassie (rpt.) (1073359). 8.10 Smurfs' Adventures (5677682) 8.35 Cartoon Critters (rpt.) (S) (1888525). 9.5 Spiderman (rpt.) (S) (2131866). 9.25 The Village by the Sea (T) (rpt.) (7767885). 9.50 Puppydog Tales (rpt.) (3358750). 10.0 Playdays (S) (7395330). 10.25 Man in a Suitcase (T) (2067750). 11.15 Resolute Bay Stories (rpt.) (9235243).
12.0 **CRICKET: SECOND TEST:** England v Pakistan (S) (90934).
1.0 **THE BROLLYS** (45104885). 1.15 A-Z of Food (24126682). 1.25 1.40 The Oprah Winfrey Show (T) (S) (8254243). 2.20 Crawshaw Paints on Holiday (rpt.) (87527408). 2.45 A Life of Knowledge (rpt.) (4677972). 3.0 News; Regional News; Weather (T) (1973798). 3.5 The Natural World (T) (rpt.) (S) (1037021). 3.55 News; Regional News; Weather (T) (3558069). 4.0 Cartoon (5707156). 4.5 Little Mouse on the Prairie (T) (S) (3567717). 4.30 Bouncing Back: The Best Bits of Johnny Ball (576). 5.0 Newsround (7336773).
5.10 **BYKER GROVE** (4503663).
5.35 **CRICKET: SECOND TEST:** England v Pakistan (S) (463446).
6.30 **THE AMAZING COLOSSAL SHOW:** New comedy series (791).
7.0 **SEVEN AGES OF MAN (T):** With Sir John Harvey-Jones (S) (2137).
7.30 **SIR* (T):** Letters to The Times between 1913 and 1919. Last in series (S) (205).
8.0 **THE STREET (T):** The team visit Pilton near Edinburgh (S) (1885).
8.30 **ONE FOOT IN THE PAST:** Castle Leslie in Co Monaghan (S) (6232).
9.0 **THE TRAVEL SHOW:** London's town-house hotels (S) (4330).
9.30 **DARK SECRET (T):** New series looking at gender issues (S) (91663).
10.0 **● HANCOCK* (T):** See Best Comedy (rpt.); followed by **Video Old Nation Shorts (T)** (84885).
10.30 **NEWSNIGHT (T)** (602601).
11.15 **CRICKET: SECOND TEST:** England v Pakistan. Highlights (S); followed by **Weather** (294866).
12.0 **GRACE UNDER FIRE** (S) (5269828).
12.25 **HOLIDAY OUTINGS** (8203422).
12.30-7.15 **THE LEARNING ZONE** 12.30-4.0 (2403335). 4.0-7.15 (14486793).

CH 4

5.0 **4-TEL ON VIEW** (80729224). 6.35 Star Street (rpt.) (2867750). 7.0 The Big Breakfast (18427). 9.0 California Dreams (rpt.) (7773446). 9.25 The Secret World of Alex Mack (rpt.) (S) (7743205). 9.55 Hangin' With Mr Cooper (T) (rpt.) (S) (6982798). 10.20 The Pink Panther (rpt.) (S) (2716663). 10.45 The Adventures of Tintin (rpt.) (S) (687576). 11.15 Biker Mice From Mars (rpt.) (S) (8709682). 11.35 Insektors (rpt.) (3624412). 11.50 Rocko's Modern Life (rpt.) (8948243).
12.0 **MORK AND MINDY** (rpt.) (78224).
12.30 **LONELY PLANET (T):** Ian Wright visits the Australian Country and Western festival in Tamworth (rpt.) (S) (83345).
1.0 **SESAME STREET** (rpt.) (9791682).
1.55 **WALES** (45163663).
2.5 **FILM* (T): Libel (1959):** Courtroom drama, starring Dirk Bogarde (764514).
4.0 **BACKDATE (T)** (S) (208).
4.30 **COUNTDOWN (T)** (S) (972).
5.0 **RICKI LAKE (T)** (S) (8900359).
5.45 **TERRYTOONS** (328682).
6.0 **EERIE INDIANA (T)** (rpt.) (137).
6.30 **BOY MEETS WORLD (T):** Shawn and Cory become journalists (S) (717).
7.0 **NEWS (T); WEATHER** (194359).
7.50 **THE SLOT** (980514).
8.0 **BLACK BAG (T)** (S) (6953).
8.30 **HOME TO ROOST (T):** Henry sets his sights on standing as a candidate for the council (rpt.) (5088).
9.0 **● SECRET HISTORY: MUTINY IN THE RAF (T):** Documentary about the airforce strikes of 1946 when airmen serving overseas rebelled against senior officers over demobilisation (S) (8427).
10.0 **FILM (T): How To Get Ahead In Advertising (1989):** Black comedy, starring Richard E. Grant (S) (563359).
11.45 **ADULT RICKI (T)** (S) (466156).
12.35 **KIDS IN THE HALL (T):** Comedy from Canada (rpt.) (S) (8220199).
1.5 **BEAVIS AND BUTT-HEAD:** Loud-mouthed animation (rpt.) (S) (4209267).
1.35 **LET THE BLOOD RUN FREE:** Dr Lovechild reveals his secret passion for Dorothea (rpt.) (8582129).
2.5-3.25 **FILM*: Washington Merry Go Round (1932):** Comedy drama, starring Lee Tracy (7523809).

ITV

For the HTV region:
6.0 **GMTV:** The News Hour. 7.0 Fun in the Sun: In Torremolinos. 8.40 Starla and the Jewel Riders. 8.55 Mighty Morphin Power Rangers (4230040).
9.25 **HALFWAY ACROSS THE GALAXY AND TURN LEFT (T)** (rpt.) (S) (7754311).
9.50 **HOPE AND GLORIA** (5180392).
10.20 **NEWS (T)** (5406750).
10.25 **HTV NEWS** (5405021).
10.30 **PEOPLE LIKE US** (rpt.) (40846392).
12.20 **HTV NEWS AND WEATHER (T)** (1883224).
12.30 **NEWS (T); WEATHER** (6881243).
12.55 **SHORTLAND STREET:** Steve's cunning plan backfires (S) (6866934).
1.25 **CORONATION STREET (T):** Derek celebrates his birthday (rpt.) (6623934).
2.0 **HOME AND AWAY (T):** Kelly is suspicious of Stephanie (S) (87534798).
2.25 **ONCE UPON A SPY** (rpt.) (3519886).
3.20 **NEWS (T)** (1980088).
3.25 **HTV NEWS** (1989359).
3.30 **THE RIDDLERS** (1512935).
3.40 **WIZADORA** (rpt.) (S) (3189040).
3.50 **MOLLY'S GANG** (rpt.) (S) (7912971).
4.5 **ANIMANIACS (T)** (rpt.) (S) (6713296).
4.20 **BLAZING DRAGONS (T):** King Allfire flies to the moon (S) (2295021).
4.45 **THE SCOOP (T)** (S) (1420595).
5.10 **A COUNTRY PRACTICE** (S); **HTV Crimestoppers** (5834359).
5.40 **NEWS (T); WEATHER** (291345).
6.0 **HOME AND AWAY (T)** (rpt.) (S) (193446).
6.25 **HTV NEWS (T); WEATHER** (178137).
6.25 **WALES TONIGHT (T)** (HTV Wales only) (785330).
7.0 **EMMERDALE (T):** Kim takes revenge on Chris (S) (9663).
7.30 **● THE BIG STORY:** See Best Health Programme (S) (359).
7.30 **A VISIT TO THE EISTEDDFOD:** Chris Segar meets Super Furry Animals (HTV Wales only) (359).
8.0 **THE BILL (T):** The Sun Hill officers investigate a vicious assault on a drunken down-and-out (8311).
8.30 **THE FREDDIE STARR SHOW (T):** With Wayne Dobson (S) (7446).
9.0 **HEARTBEAT (T):** A local grave is desecrated and the corpse's trousers stolen. Nick Berry stars (rpt.) (S) (5935)
10.0 **NEWS (T); WEATHER** (71311)
10.30 **HTV NEWS; WEATHER** (810243).
10.40 **UNBRIDLED PASSIONS:** Philip Hobbs prays for a winner at the Cheltenham Festival (617682).
10.40 **THE SHERMAN PLAYS** (HTV Wales only) (617682).
11.10 **SUMMER GETAWAYS** (S) (277514).
11.10 **THE BIG STORY** (HTV Wales only) (S) (277514).
11.40 **BODIES OF EVIDENCE:** Ben investigates a murder (rpt.) (935601).
12.35 **CUE THE MUSIC** (S) (4565809).
1.35 **NOT FADE AWAY** (S) (8691915).
2.35 **FLUX** (S) (8992809).
3.35 **LATE AND LOUD:** Heated nocturnal debates (rpt.) (S) (8469408).
4.30 **THE TIME, THE PLACE:** Studio debate (rpt.) (S) (83815).
5.0 **GRASS ROOTS** (87118).
5.30 **MORNING NEWS** (72199).

Fig. 2 TV schedules. *Source: Daily Mail*, 8 August, 1996

British television: one day's programming, four terrestrial channels. When studying television, items such as daily programme listings may be regarded as secondary source material. They may provide evidence of the importance of genres to television output.

Using the content analysis approach, you could quantify the hours given to genre material, perhaps noting the pre-eminence of some genres over others.

Scheduling and competition may be evidenced by the relative timing of programmes on different channels. Other elements which can be evaluated are the place of sport, the use of films, the choice of right programmes.

2

How to Study the Media

It may seem odd to talk about ways of studying the media, but this activity is not quite so obvious as it may seem. One does not really study the media just by reading magazines and talking about their style or the sort of articles that are in them. Nor does one do it simply by finding out facts such as newspaper circulation figures, or how television is run. Though these activities may be useful, they are not enough. What one has to do is to try different methods through which to examine various aspects of the media (not just what they put out). This chapter introduces some methods and approaches to these different aspects, which are dealt with individually in the remaining chapters.

1 SOME KEY POINTS

1.1 Process

This the subject of Chapter 3. But to make the point briefly, the aspects of the media one should look at are:

- the institutions which own, run and finance the media
- the production systems which put together the material
- the conditions under which media material is put together
- the product, or material which is produced, and the meanings in it
- the audience which makes some sort of sense of all this product
- the context in which the material is received and understood.

All this should emphasize the point that **studying the media is not just about the product**, even though it is true that this is the easiest part of the process of communication for one to get at. (Easy because one can listen to the programmes or buy the papers any time.)

1.2 Investigation

Before dealing with particular methods, it is worth realizing that one is investigating the aspects or topics just described, and making sense of what is found out. It comes down to **describing** aspects such as a particular product or a particular kind of audience, and then **interpreting** what one has described.

In a general way this is a basic method of study – **describe the item carefully, then make sense of what you have described.** What kind of sense comes out of this is another matter to be dealt with in a minute.

It is also possible to say that this **investigation focuses on the HOW and the WHY**. That is:

- WHY do things happen the way that they do?
- WHY do we have the kinds of production systems and product that we do?
- HOW do these systems work?
- HOW does the audience make sense of what it reads and sees?

These are basic questions that you can ask yourself as you carry out close examination of a magazine or of satellite broadcasting systems and companies.

1.3 Repetition and Significance

One simple fact that may help your investigations is that **anything which is repeated may well be significant.** In a sense all study and research is looking for patterns of repetition. What this means is that if you are describing ownership of the media, and this seems to repeat some characteristics across most of the media, then those characteristics are significant in some way. To take another fairly obvious example, if you study magazines for women in a certain age band and find that certain topics are repeated again and again, then these topics must be more significant than those which are not repeated. How you interpret this significance is another matter. But in this case it is fairly obvious that such repetition means that the makers of the magazines think that the topics are important, that they think they will sell the magazine, that they think the readers will like them, that whatever is said in the articles will contribute to the knowledge and opinions of the readers.

1.4 Absence and Significance

It is worth realizing that there are other reasons why the topic that you are investigating could throw up significant evidence. **What is absent may be as significant as what is present.** So, for example, the fact that there are no teenage boys' magazines like those for girls does seem significant. The fact that there is virtually no hard political news in the most popular newspapers does seem significant.

1.5 Source and Significance

There may also be significance in the source of the information that you obtain. For example, if you read a book like *Naked Hollywood* by Nicholas Kent (1991: BBC) then what you find out will be the more significant because he interviewed people who actually work in the film business, when trying to provide a picture of the working practices of the Hollywood film industry. In this case one person's informed comment is worth more than a dozen identical mere opinions.

2 PARTICULAR METHODS OF STUDY

Now that we have dealt with the main topics that we may look into, and with principles of investigation, it is appropriate to look at some specific methods which range across the media. These methods are not of course mutually exclusive. They can compliment one another. It is also possible to adapt methods to suit particular needs.

An example of this is David Buckingham's investigation of *EastEnders and its Audience* (1987). In this case he interviewed the producers; he interviewed groups of young people as audience; he described and interpreted the marketing of the programme; he conducted textual analysis of certain episodes.

See what you think of the following methods of investigation.

2.1 Content Analysis

In this approach you simply break down under headings the content of, for example, a particular programme or paper, and measure it. You may express this breakdown in terms of a percentage of the total number of pages. For instance, in a given magazine it may be that 23 per cent of it is occupied by advertisements. Such an approach can also be used to objectify what is in fact treatment, not content. For example, you could add up the number of shots in a drama programme which show the heroine in close up as compared with other female characters. You will find many more shots for the heroine. This proves that one of the reasons why we know (subconsciously) that she is meant to be a heroine is because she is given so much screen time.

The great thing about such analysis is that it stops people making generalizations such as 'there's too much violence in that thriller series'. If (and it is a big if) you measure violence in terms of the number of violent acts, as researchers have done, then you can do the counting for yourself. Find out just how many violent acts the supposedly violent programme actually contains. You could even stop-watch how long they last relative to the total programme length. It may well be that the generalization is completely wrong.

So this analysis can be used to prove or disprove snap judgements on material. Of course, it may also throw up points that you had not thought about until you saw the figures.

2.2 Image Analysis

This approach seeks to **break down the elements of a given image** (whether film shot or magazine photograph), and to **find out how the meaning in the**

GENRES X VIOLENT ACTS PER HOUR

GENRE	ACTS PER HOUR	TOTAL ACTS	TOTAL HOURS	TOTAL PROGRAMMES
Spy	9.1	90	9.9	6
Fantasy	7.7	23	3.0	2
Real-Life Cartoon	7.0	12	1.8	2
Cartoon	6.9	251	36.5	139
Avant-Garde	4.4	30	6.8	8
War	5.4	70	13.0	10
Detective	5.3	131	24.6	25
Crime	4.7	101	21.5	16
Thriller	4.7	125	26.5	23
Sci-Fi	4.6	65	13.9	10
Western	4.6	67	14.6	11
Police	3.8	68	17.7	16
Alternative Comedy	3.6	13	3.6	7
Historical Drama	3.2	188	57.8	48
Horror	2.7	17	6.3	4
Comedy	2.5	195	78.6	131
Other drama	1.6	219	133.5	115
News	1.5	148	100.9	313
Magazine	1.5	208	141.4	81
Children's	1.3	84	65.9	195
Current Affairs	0.9	37	40.9	67
Sport (contact)	0.9	13	14.8	11
Documentary	0.7	107	156.6	236
Education	0.7	19	26.7	75
Sport	0.6	23	41.7	15
Soap	0.4	26	59.5	104
Information	0.3	11	33.7	104
Music	0.3	12	36.3	40
Debate	0.3	4	12.5	24
Light Entertainment	0.3	4	13.8	22
Quiz	0.2	3	13.1	29
Game Show	0.2	3	19.0	38
Chat Show	0.1	1	7.8	13
Sport (non-contact)	0.0	4	136.0	96
Others		3	21.8	42
Total		2,375	1,412	2,078

Fig. 2.1 Content analysis table – violence

Tables such as the one shown above seem objective in their use of number, their measurement of screen behaviour. But they cannot objectify the reactions of the viewers. You could try ways of rearranging the statistics to see if they yield any other significance: rearrange programmes in terms of greatest and least number of total hours; rearrange programmes in terms of greatest and least number of total acts of violence.

image is constructed into it. In fact there is often more meaning in the image than there seemed to be at first.

There are three main elements to any image:

A The first is **where the camera was when the picture was shot**. This automatically puts us, the viewer, in a particular position relative to the objects in the image. This position may be significant because, for example, we come to realize that the camera lens is pointing at the bottle of perfume in the advertisement and not just at the scene in general.

B The second is **devices used to put the image together**. These also affect one's view of what the image means, at all levels. For instance, a modern photograph may be sepia toned in order to make it seem old-fashioned (and so give it a quality of nostalgia). The use of focus, lighting, composition, and framing are all devices which can affect our understanding of what is actually in the picture. It is rather like talking about how one says something as opposed to what one says.

C The third element is the **content of the image**, the objects that are represented within it. And content analysis can throw up some interesting points here too, and prove that we do not usually look at images with any great care. For instance, a scene from a film may show two people fighting in a room. It is apparently just a picture of two people fighting. But the paper knife behind them on a sideboard gives new meaning to the image. It suggests that something dire may be about to happen. It suggests that the fight may turn out to be more than just a brawl.

Studying the media involves looking for messages and meanings in the material. I have already said that there is a kind of assumption (which you need to test) that these meanings are there and can influence you. Meanings come through all forms of communication, not only words. In fact it is arguable that they come more powerfully through pictures because these are more like real life than words are. That is to say, looking at a picture of a person is quite like looking at the real person, whereas looking at a set of words describing that person is not the same thing at all. It is this illusion of 'being like' which is important, and which makes image analysis important. If you are able to break into the image in a methodical way, then you are breaking into an illusion. And let's face it, a great deal of media material is pictorial nowadays: comics, television, film. Even newspapers are very visual if you think about the graphic qualities of layout and the number of photographs which fill the popular tabloids (check this through content analysis).

2.3 Tables of Information

Sometimes it is possible to study the media through published information (see resources section at the back). If you read the magazine *Campaign* in your library or look at the media pages of daily newspapers (e.g. *The Guardian* on Monday), it isn't hard to find tables of items such as newspaper readership, or advertising expenditure by leading organizations. There are well-established sources of such tables which you should be able to find in your library, such as *Social Trends*.

IMAGE ANALYSIS – AN EXAMPLE

Position Signs: the camera and the viewer

- The focal point of the camera (lens centre) is on the woman's hands (on the man's naked back).
- The woman is gazing directly at us, engaging our attention, inviting our complicity in the scene.
- The point of view is slightly below her eyeline, but relatively close to the couple, as if we have caught a moment of intimacy.

Treatment Signs: devices of filming and of processing

- The image is dominated by the black and white picture of the couple embracing – the lack of colour invokes a sense of documentary still-shot realism.
- The torn margin dividing this image from the lower product section of the whole picture enhances the sense that the couple are caught in a photograph.
- The colour work for the product section enhances the contrast and is appropriate to nail varnish.
- The couple are set against a neutral white background which is improbable in terms of naturalism, but gives the image a sharp impact.
- The lighting is used to enhance a quality of naturalism in that it emphasises contours and skin texture, especially the surface of the masculine back.
- Their image is framed off like a half-body shot from a cinema frame.

Content Signs: objects in the picture and their placing

- The embracing couple invoke a sense of romance, which is of course associated with the product. They suggest relative youth, with some passion.
- Their jeans carry cultural connotations of youth, but are also culturally universal.
- The woman's swept back hair and lack of make-up suggests naturalness but also a degree of sophistication. She is actually tall enough to hang round his neck and still look over his shoulder at us. Their hands are each hidden or shown, as a kind of visual echo. Again, there is some ambivalence about age which allows the product to appeal to a fairly wide audience.
- In terms of the whole picture, there is a sense in which the image of the couple stands as a photo-story segment, as something that happens behind the present time which the product inhabits at the bottom of the picture. It is as if the picture is trying to say, this is what did happen with Nailoid, rather than this is what might happen.

Meaning

Altogether, the image endorses the idea that appearance is intrinsic to femaleness. It associates romance, even possession and power over the male, with that appearance, and with the nails in particular. The image tries to tell us that these things are 'naturally' true. The product is to be valued in terms of beliefs and attitudes which are part of the dominant ideology. The meaning is, typically, anchored by the caption.

Fig. 2.2 Image analysis – an example

So you can study the media by studying these tables. Once more it is a matter of looking for patterns of similarity or difference. It isn't hard to spot the highest and lowest circulation. But you can add factors such as comparing popular with quality press. Sometimes it is useful to turn tables of figures into graphs or bar charts so that trends and contrasts become more obvious.

On one level you may simply find out which are the most popular newspapers or whose sales are rising or dipping. But if you also look at the cover price, and then add two thirds again for popular newspapers and one and a half times again for the quality papers and multiply by the circulation figures, then you can also work out roughly how much money they are earning. The additions allow for the proportion of income derived from advertising. Having a lower total circulation doesn't necessarily mean that your profits are lower.

2.4 Sources of Information

If you are studying the media in order to see how they communicate and why, then another way of doing this is simply to go to sources of information which reveal facts which seem significant. The tables that I referred to above are a source of information. Other resources are mentioned in the Select Reading and Resource List.

But it is worth pointing out now that there are certain sources which provide useful facts and figures. Major libraries carry *Who Owns Whom*, which lists the ownership of all companies registered in the United Kingdom. This is a useful way of checking on the labyrinth of companies that own companies. For example, you will find that few local newspapers are really under local ownership. Other useful sources are the ITV and BBC yearbooks. They tell you the broad proportions of expenditure and some facts about the range of programming. Also, every company registered in Britain is bound by law to provide a copy of its yearly report on request: if you can work your way through the soft sell and the figures, you will get some idea of how much is spent on what. Otherwise, use the resource list, and good hunting!

2.5 Use of Questionnaires

Another way of studying the media is to construct and administer some questionnaires of your own. The media and their market research arms are asking us questions every day about ourselves and our reading and viewing preferences. You can do the same thing.

You can see an example of a questionnaire on the opposite page. You will notice that it tells the receiver what the questionnaire is about, asks some simple questions about the receiver as audience first, then goes on to ask questions with yes/no answers (which are easy to process), then graded questions, ending with open questions for which any answer goes. This is a useful structure to follow.

This is a questionnaire about television viewing habits in the .... area. We would be very grateful if you would answer the following questions. Please tick the appropriate box or ring the appropriate item.

1. Are you male or female? M F

2. Are you aged?—

0–15	16–20
21–30	31–40
41–60	61 +

3. Do you watch on more than 5 occasions each week any of the following channels?

ITV	CH. 4
BBC1	BBC2
Satellite channels	

4. Do you like watching any of the following types of programmes?

	YES	NO
News		
Films		
Documentaries		
Light Entertainment		
Drama		
Quiz Shows		
Thrillers		
Comedy		
Sport		
Serials (e.g. Soaps)		

5. Do you watch a News programme —

More than once a day?	
At least once a day?	
At least 3 times a week?	
Occasionally?	

6. How would you rate the following aspects of News in terms of their importance? The degree of importance is described as follows— Very Important (VI), Quite Important (QI), Not Very Important (NVI) Not Important (NI)

	VI	QI	NVI	NI
Personality of newsreader				
Range of items covered				
Items from abroad				
Having background to story				
Picture coverage				

7. What, if anything, do you *not* like about News programmes?

Fig. 2.3 Sample page of a questionnaire on media-viewing preferences

The validity of your questionnaire depends on numbers questioned as well as how tightly you have defined your audience. For example, it is useful either to ask questions of a particular age group, or of an audience (respondents) covering a range of age, occupation and both genders.

This method of media study is obviously useful for finding out things like reading habits, or opinions of programmes.

If you could also persuade your local newsagent to answer a few questions about what magazines sell best in your area, or even talk to someone in the local media about programming policy, then you would have some useful information which could also be compared with what you find out from questionnaires to the public.

Whether you fill out the questionnaire form as you conduct an interview, or whether your interviewees do this, you are in fact conducting a **survey**.

2.6 In-depth Interviews

This approach is one often used by the media themselves and by market researchers. In essence it involves a lengthy one-to-one interview with prepared questions. Such an interview could be used to elicit information perhaps from someone who is particularly expert in their field. It is also likely to be used to find out people's opinions, perhaps about an advertisement for a certain brand of perfume. It is important to select your interviewee carefully, either because they are representative of your audience, or because they are especially well informed. Then a long interview is like taking a core sample.

2.7 Focus Groups

These are a development of the above, where one talks to a selected group about a given topic in order to gain information and opinion. Again, market researchers will, for example, show the group samples of publicity material or perhaps the pilot for a programme and ask standard questions. Certainly you could select a group of women within a certain age band and show them some TV soap material in order to find out how they make sense of the material. This is a primary source of material, whereas reading someone else's research into women watching soaps would be using a secondary source.

I am not saying that these are all of the ways to study what goes on in the media. But they do represent some major methods, and cover major topics such as Institution, Product, Message and Audience. They should make even more sense as you read on through the next chapters.

REVIEW

You should have learned the following things about studying the media from this chapter.

1 KEY POINTS

1.1 One needs to look at five aspects of the media: institutions, productions systems, production conditions, product and audience.

1.2 Investigation of the media should be based on careful description of these aspects, followed by interpretation of their significance.
1.3 Repeated patterns in the content and treatment of media material are likely to be significant.
1.4 Items which are missing or not mentioned may be significant because of this.

2 METHODS OF STUDY

2.1 Content analysis tries to quantify exactly the amount and nature of material.
2.2 Image analysis breaks into the meaning of visual material through careful description of where the camera is placed, of technical and other devices which contribute to the treatment of the image and through careful observation of elements of image content in relation to one another.
2.3 Analysis of tables of information available through various sources.
2.4 Going to reference sources of information.
2.5 Using questionnaires to investigate audience knowledge and attitudes in particular.
2.6 Using in-depth interviews to investigate attitudes and knowledge in personal detail.
2.7 Using focus groups to investigate opinions and attitudes of a cross-section of the audience at one time.

Activity (2)

This activity is concerned with the methodology of media study. The idea is to take a media text or product and try at least three different ways of 'getting into' it.

This approach could provide reinforcement for making certain judgements about the product. It will certainly provide a range of evidence. It should make it clear that there is no perfectly correct way to study the media.

Although you are dealing with product in particular, you could equally well try to find out about the institution and marketing behind the product, or about the nature of the audience which consumes the product.

Take one example of a comic or of a magazine such as *More*.
You will probably find it useful to have three copies.
You may want to work with other people on this one.

MAKE A CONTENT ANALYSIS, in which you measure the percentage of certain kinds of content, or you calculate the frequency with which certain kinds of treatment appear.
For example, you could –

- measure the proportion of pages given to illustrations or to advertisements or to certain topics
- add up the number of items which appear repeatedly in stories or articles or pictures (such as acts of various kinds of violence).

CONDUCT A FOCUS GROUP SESSION, in which you get the opinions and reactions of the group to the material and to particular questions you ask.
For example, you could –

- ask them to talk about how the stories in a comic connect with other things they have viewed or read
- ask them to explain why they are attracted to or repelled by certain characters, and why.

ADMINISTER A QUESTIONNAIRE, in which you also concentrate on reactions to the comic or magazine. You would have to make sure that your respondents had seen/read even just one part of your material.
For example, you could –

- construct questions to find out about what they liked best, what they remembered most, and what they feel the sample is mainly about in terms of meaning.

You should compare the results of these three approaches to the product to find out, for instance, whether or not what attracted people to characters in the focus group was the same as what they liked best as shown by the questionnaire, and was something which scored 'highly' in the content analysis.

M
E
G
A
M
I

3

A Basis for Media Studies

This chapter describes the backbone stiffening this book. I have already suggested some reasons why we should study the media in the first place. I have described some methods of study. Now I want to present some ideas and terms which I believe are central to Communication/Media Studies. These ideas help us interpret our findings. They help make sense of how and why the media communicate in the way that they do.

1 THE PROCESS OF COMMUNICATION

All acts of communication are a process. This process includes a source, a message and a receiver of the message. **When the media communicate with their audiences, there is a process going on.**

From one point of view, we may study the media by fastening on parts (or factors) of the process and seeing how they affect the creation and understanding of messages. It is important to remember that **any one of these factors within any communication process will affect the content and treatment of its messages.** So it is also true of the media.

In describing the process of communication in the mass media Stuart Hall (1980) talks about 'a structure produced and sustained through the articulation of linked but distinctive moments – production, circulation, distribution/consumption, reproduction.' It is these 'moments', and others, which we are going to look at.

If process includes the idea of message, then the idea of message includes the notion of meaning. It is arguable that the meaning is the message. Again it is Hall who says, 'if no meaning is taken there can be no consumption.' But we certainly do consume media products, and in doing so make meanings from what we consume. I say 'make' meanings because although some people talk about taking meanings from texts, it isn't really true that we simply swallow the meanings from a magazine like pills from a bottle. As I will explain later, some of the meanings are to an extent determined by the producer, but it is also true that we have at least some freedom to make sense of media material in the way that we want to. The nature and degree of that freedom is shaped to a large extent by determining factors in the process of communication through the media.

So let's look at some of these basic factors and apply them to the media. This will help you see how the remaining chapters in this book fit in.

1 .1 Source

Media institutions are the source of many types of message that we receive. These media may be responding to events and opinions in society at large, but they are at the same time the composers and initiators of the communication. So we need to look at their characteristics, at the way they operate, at their reason for communicating, in order to understand how and why the messages are shaped.

1.2 Need

All communication answers some need in the sender or the receiver. We have just referred to the source: we will also look at the receivers or audience and discuss whether communication satisfies their needs within its overall process.

1.3 Encoding

All messages have to be put together (encoded) in some form of communication. How messages are put together is bound to affect how they are understood. A news item on television is not the same thing as it is on radio, even though both are broadcast media and tend to cover the same topics. For example, if you have pictures of an event they provide an immediate sense of action and background, of being there. (This is certainly not to say that television is superior to radio in any way!)

1.4 Message Content/Message Treatment

All messages can be defined in terms of what they say and how they say it. Every message conveys some kind of meaning to the recipient. But how the message is handled has an enormous effect on how it is understood, and indeed on what is understood. In a drama about the injustice of war, this message comes across in one way if the play relies on a narrator to make comment, and in quite another if recordings of authentic war action and war speeches are used. The basic message is the same but the different treatment means they do not have quite the same meaning.

1.5 Decoding

How one 'unpacks' a piece of communication to get at its meaning obviously affects the kind of meaning that one gets. To make a basic point, most of us decode communication through the filter of our own experience and indeed our prejudices. If we stick with the example of the war drama, then it should be fairly obvious that a member of the armed forces will probably decode that

A

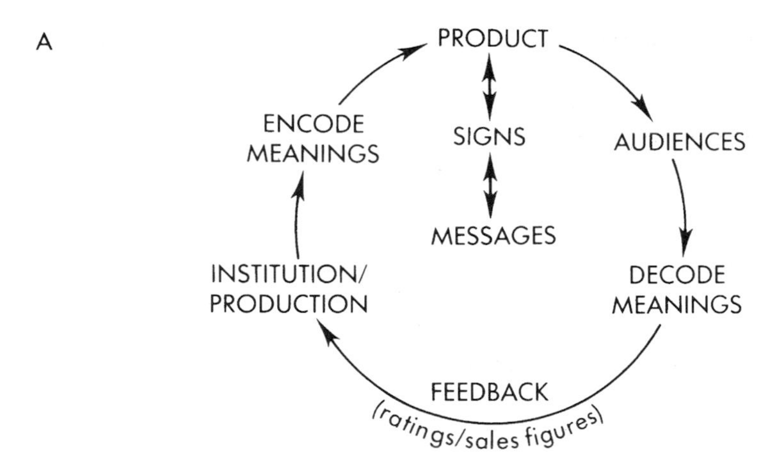

What the producers mean to say and what the audience think they mean, may not be quite the same thing.

B

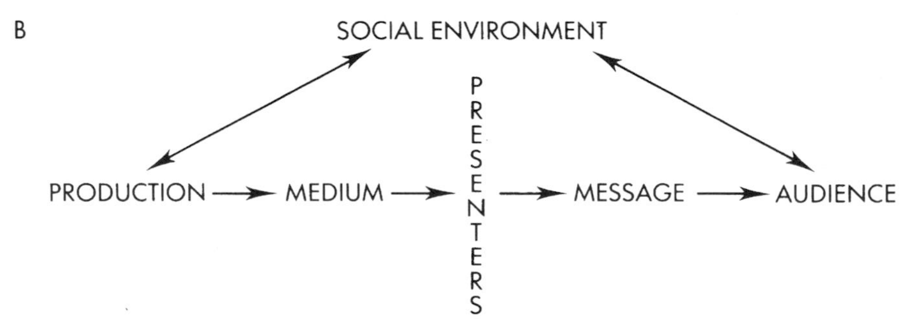

The audience is directly in touch with its social environment, but also receives messages about it indirectly, often through influential media personalities (presenters).

Fig. 3.1 Two models of communication through the media

drama rather differently from a committed pacifist. The media communicators are well aware of our tendency to read what we want to read into communication. So, intentionally or not, many examples of media communication are structured in such a way that they push us into getting the message originally intended (see later comments on preferred readings).

1.6 Context

All communication is carried on in some kind of physical or social context. The context always affects how the communication is understood, and maybe how it is put together in the first place. On a simple level, one of the reasons why tabloid newspapers have the format they do is that they are relatively easy to open and to manage in buses and canteens – that makes them more attractive to use. But more important might be the example of television, which is received in the context of our own homes. It has to compete with family chat and distractions. Children's programming is organized around the assumption that the audience will be home at a certain time to watch. The volume on advertisements is turned up so that they will be at least heard if people start talking between programmes or even go out to make a cup of tea.

In these cases we are talking about the contexts in which we receive or consume the communication.

One can also take account of the **context of production** which is to do with the circumstances in which the media material is produced. In this case it matters, for example, that a film is made as a collaborative enterprise by a large group of people, and is not shot in narrative sequence. Whereas at the other extreme a novel is in its creation a solitary enterprise, probably written in sequence.

There is also the **cultural context**, which affects both producer and audience in terms of how the product is conceived and understood. There is a British comedy series called *Absolutely Fabulous*, which has been sold to the US. The US version is being transformed to take account of nuances of British class and taste which are not understood on the other side of the Atlantic.

1 .7 Feedback

All communication will get a response in some form, however delayed. Record and magazine producers appreciate feedback in the form of good sales. What is more interesting is that they, along with other media makers, do not encourage rapid feedback in terms of comment on the product. And of course it is difficult to offer such feedback compared with the example of face-to-face communication. But whatever feedback is received in whatever form, it does affect the communication. When the British soap *Brookside* started out, the producer sought heightened realism by using some (authentic) bad language. But the feedback from the audience was poor – too much reality was hard to take. Ratings were affected. So the scripts were changed.

1 .8 More on Process

The idea of process is actually a little more complicated than simply a list of parts, as represented by this set of topics. One should understand that **the word process refers to something which is active, dynamic and which has continuity**. Communication through the media is not something like a little set of packages being delivered. It is rather more like a continuous flow. This process involves the different media interacting with each other, as when the television news editors read the early morning papers to see what their version of the news is. It involves the media interacting with society at large, as when the soap serials pick up current social issues and events and weave them into their storylines.

So the idea of process can lead one to look at the sequence of events in which meanings are created and received by the audience, with reference to particular programmes or publications. But we should also look at how the audiences are connected with the media in terms of their responses to these products.

Process requires us to look at the wider social context in which the meanings of the messages are created and understood by the audience. For example, analysis of young women's magazines provides ready evidence of powerful messages about the importance of appearance and image to young women. But these messages do not only appear in such magazines. Apart from other media, they are also generated through peer group discussion, which in itself will affect how these messages are understood.

2 TEXT AND TEXTUAL ANALYSIS

Another basic way of looking into media material is to regard it all as TEXT of one sort or another. This approach regards comics, television programmes, and films as being texts, as much as the material that relies on written forms of communication. **A text can be read**. It will be organized or structured in particular ways, just as the text of this book has been written to a structure. The idea is that if you can work out what the organizing principle is, and if you can analyse the way things are 'said', then you have a good chance of working out what the material means.

Textual analysis might use the methods of study described in the previous chapter. It would also look for structure. It would treat all media material, visual or otherwise, as a kind of 'book', with meanings to be read into it. Indeed, one can only make sense of a text because it operates within a system of meanings that we share in our culture. This system in turn depends on signs and codes, which are discussed a few sections further on in this chapter.

You will find that there is a close relationship between concepts such as 'text', 'sign', 'structure', 'narrative', 'code' and 'convention'. In particular, textual analysis, structural analysis, and semiotic analysis, though dealt with separately, all involve understanding of similar concepts, and are all ways of getting to the same thing – how the text is put together, how we read meanings into it. In fact I would suggest that they are all aspects of the same thing. **Semiotics** concentrates on the building blocks of the text to get to meanings – the words and the elements of pictures. **Structuralism** looks for organizing principles and at whole sections of text – chunks of narrative, or the *mise en scène* of a film shot. Text is a more all-embracing term. But in truth it is difficult to deal with one of these approaches to media material without moving on to one of the others.

With regard to what these meanings are, when looking at the media it is difficult not to end up talking about aspects of society and politics. The treatment of old people in sitcoms or the reporting of police handling of a travellers' convoy will both say something about social groups and about power. John Fiske (1990) says that 'every text and every reading has a social and therefore political dimension, which is to be found partly in the structure of the text itself and partly in the relation of the reading subject to that text.'

So again he is saying that the sense we make of a text is both in the way that the text is put together and in the way that we choose to make sense of it.

In terms of reading meanings into a text, it is also possible to define two kinds of text. A **closed text** is one where the way it is put together closes down the range of possible meanings one may get from it. An **open text** is the opposite – it is possible to make sense of that text in a number of different ways. Even with an open text there are a lot of cultural pressures on us and assumptions that we make, which may limit our interpretation.

3 STRUCTURAL ANALYSIS/STRUCTURALISM

In terms of words, of written language, the structure that we get to by analysing a text is essentially that of grammar. This means that we are also talking about conventions or rules, which are organizing principles. These will come up again under the heading of semiotic analysis. Other organizing principles within language are the rules of spelling and the rules of word order, or syntax.

But all 'languages' and all media can be analysed for their structures. The proposition is that all texts have an underlying system of elements and rules which help produce the meaning of a text. Genres (see Chapter 5) would be a particularly recognizable example of this. This principle of structure has caused critics to look for basic elements in a text – types of character or patterns of story line – and then look for principles by which these are put together. Strictly, this is about looking for how the meaning is put into the text, not what that meaning is.

This approach also has problems, it has to be said. For example, it seems attractive to suppose that many stories use the element of the 'villain' character – from the witch in *Hansel and Gretel* to those various Asian, eastern European and Russian villains in Bond stories. The trouble is that the meaning of 'villain' is not necessarily 'written into' the structure of the text. It is also constructed in the head of the reader/viewer. With a given story, different cultures might read different characters as villains. So they would not see the text as being structured in quite the same ways.

Graeme Turner in *British Cultural Studies* (1992) talks about a division between structuralists and culturalists. He suggests that structuralism at least has been more interested in form and structure and producing general meanings and principles, than in coping with details of culture. There is a cultural tradition which is almost humanist and which resists the idea of labelling things too precisely. I suggest to you that as a student of the media you try different approaches and that you **resist the idea that there is any one correct way to make sense of the media**. Often the differences between academic approaches say more about academic culture than they do about 'correctness'. I have found that few ideas are mutually exclusive, and that many can be accommodated within one another – semiotic analysis within process, for example.

Having said that, the idea of structure does still lead us to two more notions, which are very useful.

3.1 Binary Oppositions

Binary oppositions are opposing concepts which one reads into the text usually through contrasting sets of words or of pictures.

The most basic oppositions are to do with good and evil or with male and female. One can then find words or picture elements lined up on one side or the other, to underline the opposition, and of course to suggest approval or disapproval of one element or the other. Males are tough, hard, reasonable; females are pliant, soft, emotional. Villains are filmed in shadows, in dark clothes, with unshaven faces; heroes are clean-cut, in pale clothes, in light.

Although I suggest elsewhere that there are more than two sides to every story (especially a news story), it does seem deeply ingrained in our culture that

Fig. 3.2 From *EastEnders* – David and Carol argue again

The still of David and Carol from *EastEnders* 'facing each other out' in another argument, very much represents the pervasiveness of oppositions in narrative in general and in soaps in particular. They are even framed with a space between them but are clearly facing up to one another. You would find it informative to take any soap, list in pairs those who are in conflict, and list reasons why they are in conflict. Patterns will emerge. For example parallelism emerges – the subplot. If one couple is 'slugging it out' in a domestic conflict, it is likely that there will be another pair echoing this opposition.

we should think in these opposing ways. Many texts do have this structure built into them. Many stories are based on conflict, and the easiest conflict to set up is that between two people or two views. There may be more than one set of oppositions in a story. To describe this structure is to describe how the text is put together. One has found a pattern. But to make sense of the text one has to explain what the opposition means. Usually that meaning is about the positive and the negative: one of the opposing elements being valued in terms of 'right' or 'good' or 'attractive'; the other as 'wrong', 'bad', 'unattractive'. In fact one is into what is valued and what is not – aspects of ideology.

3.2 Narrative Structure

Indeed it is possible to see the organisation of the narrative (defined roughly as the storyline and the drama) in terms of opposing building blocks. More is said about narrative in Chapter 5. But, at this stage, you still need to understand that describing and interpreting the structure of the narrative is another kind of structural analysis. For example, what is called **mainstream narrative** or the **classic realist text** has a developmental structure. This is your average story in most media, where the plot develops from some initial problem or conflict, through various difficulties to some neat ending where everything is sorted out. Analysis of such narrative structures not only leads to understanding of how we come to see that text meaning what it does, but is also likely to help explain how we, as readers or viewers, are positioned in relation to the text. For example, autobiography depends on us being privileged to see into the mind of the storyteller and to see things through his or her mind.

Looking for a structure in the narrative of a text leads to more than just a description of how the 'machinery' works. It helps explain how we understand a text. It helps us understand that what we think a text means is firstly more complicated than it appears on the surface, and secondly is not a matter of chance. So again the idea of structures forms a basis for media studies.

4 MEANINGS

Process includes the notion of **ideas about people and their beliefs being kept in circulation through the media**. This brings us back to the central importance of meaning. These messages and meanings are both **overt** (the most obvious and apparent) and **covert** (more or less concealed and implied).

We have suggested then that a basis for Media Studies is the idea of process, and that **the core of process is meaning.** In terms of meaning it is also basic that the communication is about what is normal, what is entertainment, what is news, what is important, what is valued, what we should believe in. These are the meanings that we may look for in the process. And of all of these it is **the value messages that are most important** because what we value is what we live by. So I suggest that it is basic to study of the media that we look for what is valued, why it is valued, and what effects these values may have on us, the audience.

Fig. 3.3 From ITN titles sequence – Big Ben

This very familiar image from the main news titles sequence has strong meanings tied to culture, myth and ideology. The power and symbolism of the image, as the shot homes in on the tower from above, is what caused it to be used in the first place. The Big Ben tower and clock stands for the secure centre of things, being associated with Parliament itself. The programme is about weighty matters of news and it comes from there – the centre of our political process. Big Ben is a cultural icon of Englishness, as the Eiffel Tower is for France. Its image unlocks the central precepts of our dominant ideology – of course we believe in Parliament, in its democracy, in the natural rightness of the fact that London is the centre of our universe, of the fact that our world should be run from the building below the clock. In fact, all of this is really about the myth of being British and, among other things, about that part of the myth which helps us to see ourselves as being somehow important in the world simply because of our Parliament. You take it from there!

To study the media is to study meanings – where they come from, what they are, how far they are intentional, how they are built into media material, and how they are incorporated into our own thinking.

For example one idea which is very much valued is that of mothers and motherhood. Pictures in advertisements, storylines in soaps, and feature articles in magazines are some examples of media product where you are likely to see/read about motherhood in an approving light. In fact approval of motherhood is built into our ideology, our way of looking at the world. And to this extent you need to understand that **it is basic to media studies that you**

will keep coming upon ideology whatever concept you start with, or whatever analytical approach you take.

You can start with Genre or Representation, for instance. But still the value messages within examples of these are the values within our ideology. Keep reading, and when you come to such ideas in this book then you will see the truth of this statement. As you read on, you should also understand better how the main ideas that are basic to Media Studies all connect. They are like the different-coloured threads in a garment which are sewn together to make the whole thing.

5 CULTURE

The study of media is almost inevitably the study of culture. As a basis for Media Studies one needs to recognize that the texts which we discuss come out of our society and our culture. This means that they will contain references and meanings which are particular to our culture. For example we can use phrases like 'working class' that will have some meaning for most people in Britain, but little meaning in the USA. This is not to say that our media products are incomprehensible in other cultures. But it does mean that what is comprehended may be different: there will be different interpretations. In particular, the value messages referred to above could also be described as cultural values.

There are courses in cultural studies, aspects of which are the same as media studies. Other parts of these courses might seem to have more to do with sociological analysis. For example one could look at teenage girls and the magazines that they read, both through analysis of the magazines, and through surveys of how those magazines become incorporated within their social lives.

In media studies you need to be very aware of the idea of culture because it does keep coming in, as one makes sense of media products in particular. For example if you were to examine regional media in Scotland then this would also become a cultural study. This would be because you would come across different newspapers and TV programmes that talked about subjects and made assumptions that are specific to Scotland and to the way that people think in that part of Britain. These differences may not be huge but they are there. You only have to think of how Scottish people are represented in sitcoms to see that we are aware of those differences, and so are aware of something called culture, and cultural values.

6 SIGNS AND MEANINGS

6.1 Signs

We have said that the process of communication through the media helps create vital meanings in our heads. We have said that these meanings come to define how we see the world. Even on a simple level, our idea of what is entertainment can be so defined. Television defines all of its output, but especially comedy, as entertainment, hence the pat preview phrases: 'that's Friday night's entertainment to look forward to on BBC1.' We come to see it as natural that television should be about entertainment, and perhaps feel that anything that is 'heavy' is somehow unnatural. That is the meaning that has been created. But it is anything but natural. **Communication is not natural. It is learned, both in its encoding and its decoding.**

The meaning in media communication is signalled to us. It can be taken as basic to all communication study that this signalling and signing does take place in all sorts of ways. Headlines signal to us what we should understand is priority news. News photos make signs about who and what is to be considered important.

There are specific signs which do this signalling in whatever form of communication is under discussion. The media encompass all forms: the spoken word on radio, the photographic image in magazines, the printed word in newspapers, and combinations of all these in television. So it is a fact that recognition and analysis of specific signs is crucial to the understanding of these messages and their meanings, e.g. the word 'disaster' in the headline and a close-up of a grief-stricken face below this.

The meanings of these signs are learned. They do not 'naturally' belong to the sign. This separation between sign and meaning is important because it explains why people may see the same signs or material as meaning different things. Formally speaking, the sign is called the **signifier**, whereas each one of a number of possible meanings is called the **signified**, and the meaning which the receiver gives to the sign is called the **signification**. The word 'disaster' on the page is a signifier. It could mean many things. There could be a whole range of signifieds in the reader's head, including various types of possible disaster. Usually other signs (or words and pictures) will help pin down the intended meaning. In this case there could be a picture of a smashed aircraft beneath this single-word banner headline, and it would be clear that an air crash is the signification.

Individual signs may signal strongly to us. But in the end it is always the collection of signs which add up to the complete **meaning in a message.** Ultimately one has to read the complete sentence, the paragraph, or the entire news article, in order to make sense of it.

6.2 Codes and Conventions

Collections of signs in specific forms – such as speaking, writing and pictures – are known as codes. These codes are also defined through their conventions – or the unwritten rules about how the code hangs together, how it is used, how it will be understood. All examples of communication are bound by conventions. We learn these rules as part of our process of socialization. Precisely because of this we are often not aware that these rules exist. Once more they seem so natural that they become invisible. To study communication is to try to make the conventions (and the codes) visible.

If we consider **primary codes such as writing or visuals,** then some of the conventions are very basic. For example, there is one which says that we write from left to right starting at the top of a sheet. This is not what happens in some other cultures! Some of the rules of visual codes are just as basic – we learn to look for 'important' items in the middle of the frame, rather than at the edge: we have learned to accept, through documentary photography, that the scene may be cut off at the edges by the frame, so that everything is not neatly contained within it. You will see that these conventions are shared by the producers and the audience. Communication depends on agreeing on these rules so that encoding and decoding are pretty well matched and at least the main meanings are put across successfully. These rules may change over a period of time as creators of communication experiment and invent. Even spelling, for instance, may change, in spite of the efforts of the dictionary makers. The rules of visual narrative have changed as people become more attuned to picture sequences. It is no longer necessary to use some conventional device such as a wave effect and dissolve to represent a shift back in time. People can take a straight cut, and pick up the threads.

Do also remember that all the visual media draw on such primary codes as non-verbal communication (NVC). We understand these too through conventions which organize the use of these signs. We learn how to use NVC from the screen as much as from real life. Part of the meaning of film, television and photographic material comes from our being able to decode these signs. A combination of codes and conventions can establish meaning such as a cut between close-ups of faces of two people in a group apparently looking at one another, which means that there is some connection or relationship between the two.

6.3 Secondary Codes

There may be codes within the codes which we also learn to understand. These are secondary codes. Students of the media must learn to recognize the signs, to unravel the codes, to get at the meanings. For example, there is a secondary code of television news. An instance of this would be those special signs which 'mean' the credibility and truthfulness of the newsreaders. The signs involved include the face on close shots, the formal dress, the generally serious expressions and style of speech, the face-to-camera half-body shot usually of the reader sitting at a desk.

These secondary codes also operate through conventions. Some of the rules are quite practical, such as the use of columns in newspapers and the assumption that we read down one column and then go to the top of the next one. These are the rules of the production of a given format. You learn to use these when, for example, your teacher asks you to produce a leaflet or a booklet. But there are secondary codes which are to do with rules particular to a type of product. For example, women's magazines include special signs and conventions which define how we expect the models in the photographs to pose. These are very different from the way we expect people to pose in family snapshots. **All genres or modes of realism are also based on secondary codes with their own rules.** Categories of television material are likely to have their own secondary codes. For example, we are now well used to seeing an immediate rerun of a goal scored in a screening of a football match.

(You can find out more about these concepts of sign, code and the creation of meaning from books such as *An Introduction to Communication Studies* by John Fiske and *Understanding News* by John Hartley. You may also find *More Than Words: An Introduction to Communication Studies*, by Richard Dimbleby and Graeme Burton, useful for chapters 1 and 5 on basic theory and on the media.)

So the **media are full of codes and signs working away within the overall process**, all telling us how to understand the material. These signs and codes are held together by rules called conventions. To take one more example, consider the cover of a typical magazine for women (and the word typical should alert you to the existence of conventions). This cover may well include elements such as a smiling female face (non-verbal code), that face centre frame of the cover (pictorial code), the head shot with titles of articles around it (magazine secondary code). All of these signs combined by convention add up to meanings such as: *this is a magazine for women*; *pay attention to this cover*; *be happy*; *read on inside*; *this is what women are/should be interested in.*

So another basis for your media study, apart from approaching the material as a text, and the topics through examination of process and analysis for meaning, is to break it up into signs, codes and conventions. You can look for these. You can use what you find out to explain how and why communication takes place and why this matters.

7 CONSTRUCTION OF MEANING

Meaning is constructed into the material through signs and conventions in particular. This construction is something which the audience does, as well as the producers. The meaning is not something like a parcel which is wrapped and passed on to the audience. It is more like a set of blueprints for a structure which the media producers expect the audience to follow and which they design quite carefully. But still the audience may build something else from this design. The audience is not a passive receptacle for what the media has on offer. **To study the media is to study how these meanings are constructed by both producers and audience.**

Take the example of a certain kind of television light entertainment show which features some 'star' personality. Consider the likely beginning of this show. The star descends a grand set of stairs back of the stage, accompanied by triumphant orchestral music. There is canned applause or applause from a studio audience. The lighting is dramatic with a key light on the performer. The rest of the set dressing is theatrical. The star's clothes appear to be elaborate and expensive. The camera is on a crane and a dolly (small truck), so that it can follow the performer and can carry out elaborate movements, changing height and angle. All these signs, which are conventional within the given secondary code and which are used together by convention, send complex messages or meanings to us: 'this is an important person descending as if from heaven'; 'this is a production worth watching because money appears to have been lavished on it'; 'this is the beginning of a show'; 'this person has star status'; 'we are expected to treat this as entertainment and to enjoy it'; 'we should share in the approval and pleasure of the audience'; and so on.

8 THEORIES ABOUT MASS MEDIA IN RELATION TO SOCIETY

As a broad basis for looking at the media it is worth summarizing a number of general theories which try to describe how the media operate in relation to society as a whole. You may wish to discuss which of these theories seems to fit the set-up that we have in Britain. More is said about these in Dennis McQuail, *Mass Communication Theory*, in particular.

Authoritarian Theory sees the media as a means of communicating authority views. This theory also suggests that they should be used to produce a consensus or agreed way of looking at things in the society as a whole.

Free Press Theory sees the media organized so that anyone can say anything at any time. All these views will, it suggests, balance each other out in the end.

Social Responsibility Theory sees the media as working to an ideal of objectivity, and on the basis of having a sense of obligation to society as whole. Media operating this system would offer diverse views, but would draw a line somewhere – not encouraging, for example, violence or criminality.

Soviet Media Theory sees the media as being used for socialization, for creating public opinions, for education. This theory suggests that the media should be in the control of the working class.

Developmental Media Theory sees the media as being there to develop national culture and language. This theory also suggests that the media should carry out tasks of social and educational development within the framework of some national policy. This view of what the media could be like and how they should work is associated with what does and could happen in Third World countries which are trying to develop their economies and resist the cultural influence of the developed nations (e.g. not buying hours of American television!).

Democratic Participant Theory sees the media as operating through a great variety of types of media organization. This theory suggests that there should not be any centralized bureaucratic control of the media. It suggests that the media should be organized to encourage the rights of minorities and individuals to have access to the media and to use them.

You should cross-refer these ideas with the description of functions in the chapter on institutions.

Such theories are based both on an interpretation of how the media work in various societies, and on ideas about how they might work. Of course one's opinion about each theory partly depends on ideological perspectives. For instance, if you run the Chinese state press then you don't have much problem with some authoritarian perspective on how the media should be run. You will believe that the state authorities should control the press for the good of the people. On the other hand if you run the CBS network in the US then the Chinese view would not be OK, and you would likely subscribe to a view which combines freedom with social responsibility.

It is useful to be able to take a broad view of the media in a given country and to describe them in principle. It is likely that you will tend to approve of one system or another. But still you need to be careful about instant disapproval of some systems, because it might say something about your ideological blinkers. As a media student you need to be aware of ideology and its workings, and try not to be trapped within it.

In particular it is useful to be able to look at examples other than our own in Britain, to be able to stand back from what we take for granted. In Holland, for example, the media are more libertarian than ours are, with less controls in their broadcasting acts and in their laws. Access to broadcasting is more firmly democratic in that 'broadcasting organizations' are defined by the size of their membership and have an equally well-defined right to certain amounts of air time.

We might like to think of ourselves as belonging to the democratic participant model. We might take a pluralist view of our media, believing that we have many choices of channel or of magazine, with a range of views. But I am about to point out that our media organizations are not that different in the ways that they are set up and the ways that they work. Perhaps we don't have that great a choice of material when you consider how much of it falls under genre headings – all the same type done in much the same way. And maybe there is more of a centralized control of the media than we would like to think. The ITC is a kind of bureaucracy which has a great influence on commercial television. The Government interferes in various ways that you will read about. A few newspaper proprietors control the output of most of our press. This isn't direct state bureaucratic control. But neither is it a great example of freedom, especially if one expects to see recognition of the rights of minorities or good access to the media for ourselves, the audience.

The advantage of thinking about the whole media system and of not taking our own for granted is that it is then possible to think about different ways of doing things. Who should run things? How should they best be run? How could we pay for our media? It isn't inevitable that our media should be dominated by financial interests in terms of control and of finance. For example we could decide that we want alternative views in different kinds of national newspapers – views which don't fit in with those of advertisers who substantially pay for our mainstream press. We could decide that all commercial newspapers have to pay a levy to fund an alternative national press.

Such ideas, together with the description in this chapter of some key concepts, should give you a basis for studying the media. What follows now develops ideas about how the media operate and more importantly how we make sense of them in terms of Institution, Product and Audience.

REVIEW

You should have learned the following things as a basis for Media Studies.

1 PROCESS

- all communication is a process. The media are no exception to this. There are key factors in the process of communication through the media which one may look at. These are source, need, encoding, message content and treatment, decoding, context and feedback.
- process also involves looking at how meanings are created, and taking account of the social/cultural context in which media communication takes place.

2 TEXTUAL ANALYSIS

- all media material, visual or otherwise, may be seen as a text to be read.
- this reading may involve a variety of approaches to get at the meanings in a text, including structural analysis.
- there are open and closed texts.

3 STRUCTURAL ANALYSIS

- all media texts have some organizing principles or structures within them.
- two useful examples of these are binary oppositions and narrative structures.

4 MEANINGS

- meanings are in messages, which may be overt or covert.
- meanings and messages are often about values.

5 CULTURE

- texts and their meanings are very much governed by the values of their given culture.

6 SIGNS AND MEANINGS

6.1 The meanings in media communication are signalled to us through a variety of signs. We need to identify these in order to get at the meanings.

6.2 These signs are organized into codes covering words and pictures. How they are organized, put together and understood depends on rules or conventions which we also need to recognize.

6.3 There are codes within the general codes of speaking, writing and pictures. These special codes are called secondary codes. They have their own rules. They help organize categories of media material such as genre, and treatment of media material, for example in terms of realism.

7 CONSTRUCTION OF MEANING

- meanings don't just happen, they are built into media material, intentionally or otherwise.

8 THEORIES: MEDIA AND SOCIETY

- there are a number of general theories about how the media do or should operate in relation to the society of which they are a part. These may be summarized as follows: the Authoritarian Theory, the Free Press Theory, the Social Responsibility Theory, the Soviet Media Theory, the Developmental Media Theory and the Democratic Participant Theory.

- these theories should be cross-referred to ideas about media functions.

Activity (3)

This activity is concerned with the basis for media studies described in this chapter.

So it asks you to take a text and look at it through three different ways of understanding the media – indeed, any kind of communication. These approaches see the text as part of a process, as a combination of signs, as something with structure. In each case one may draw meanings from the process, signification or structure. The meanings can be about the text, the product. But they can also be about the institution behind the text and the audience which consumes the text.

Take an example of a TV programme which is current and aimed at a younger audience. At the time of writing this might be *Hollyoaks* or *Heartbreak High*.

ANSWER THE FOLLOWING QUESTIONS about the process of which this programme is a part by reference to it and to its credits.

- Who is making this programme and why?
- What are the characteristics of this type of programme, and why might this matter?
- Describe the audience for the programme, and say how you know this.
- What are similar types of programme, and how may the programme relate to the life style of the audience?

MAKE AN IMAGE ANALYSIS OF ONE SHOT/SCENE, probably from a freeze frame.

- What are the image position signs which affect one's understanding of the scene?
- What are the image treatment signs which affect one's understanding of the characters, or one's sense of realism?
- What is there about the placing and choice of people and objects in the frame which also affects one's sense of realism or understanding of the characters?

MAKE A LIST OF CHARACTERS OR IDEAS WHICH ARE OPPOSED TO ONE ANOTHER in the programme.

- How does this help define what the story is meant to be about?
- Can you also find any triangular relationships in the story, and comment on these?
- Does the end of this story/episode somehow follow on from the opening?
- Would you say that the end (or the bits before commercial breaks) are contrived in any way?

Having done all this, it would be useful to note down those points which came out more than once through these three approaches, and those which only came out because of a particular approach. This will help you confirm that there are meanings about text and its production to be brought out, and will help you make decisions in the future about which approach you might choose to use when making some media analysis.

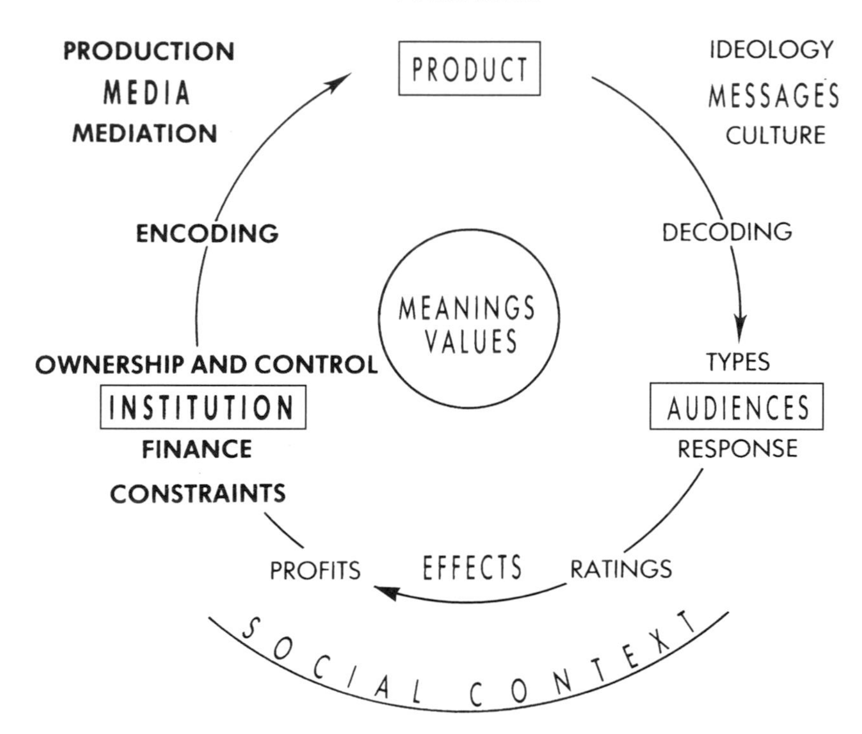
REPRESENTATION
INFORMATION AND PERSUASION
GENRES
REALISM
NARRATIVES
PRODUCTION
MEDIA
MEDIATION
PRODUCT
IDEOLOGY
MESSAGES
CULTURE
ENCODING
DECODING
MEANINGS
VALUES
OWNERSHIP AND CONTROL
INSTITUTION
FINANCE
CONSTRAINTS
TYPES
AUDIENCES
RESPONSE
PROFITS
EFFECTS
RATINGS
SOCIAL CONTEXT

4

Institutions as Source

Where Does It All Start?

In the last chapter I pointed out that the process of communication is one which ultimately has no boundaries. One thing always connects with another. For example, the people who make media material are also in a sense part of their audience. They are also members of society. They read newspapers and watch television like everyone else.

However, it is also fair to simplify the idea of process somewhat and look at it in terms of where the messages (in magazines or programmes) start, as well as where they end up. The source of the communication shapes the message.

In this chapter we will look at the nature of the institutions and their production systems, in order to see how this may affect their output. The phrase 'nature of institutions' covers how they are organized as well as what their values and operating principles are. The term institutions includes a public broadcasting organization like the BBC and the other, commercial organizations. In particular we are looking at the largest organizations operating in any one of the given media (see Figure 4.1 on page 45). This means that we are looking at a pattern in which, in each case, about five companies own about 70 per cent of their respective media.

1 DOMINANT CHARACTERISTICS OF MEDIA INSTITUTIONS

1.1 Monopoly

In the first place we can say that these tend towards monopoly. In no case does one organization have an absolute monopoly. But the **domination by a few companies is sufficient to raise doubts about choice and accountability**, as it would in the case of any monopoly. This situation could lead to what is called a cartel – where a few companies make an unofficial arrangement to carve up their particular industry for their own profit and convenience. For example, it has been proposed that the major ITV companies are such a cartel because they dominate programme purchase and production, and the other 11 companies virtually have to go along with what they want. The major companies are Carlton, Granada, Yorkshire/Tyne Tees, Meridian, LWT.

1.2 Size

All these major organizations are very large. They employ many people and have huge turnovers of cash. It follows that they are well equipped to produce and pay for expensive media material. It also follows that they are the better placed to monopolize in their particular field. It cost Eddie Shah £15 million to equip and launch a new national daily tabloid newspaper, *Today*, in 1987. He still failed to attract enough advertising to make it viable, and *Today* ended up being taken over by Rupert Murdoch's News Corporation. NewsCorp closed down the paper in 1995 because even with a readership of nearly a million it still wasn't making enough money.

1.3 Vertical Integration

These institutions are vertically integrated. That is to say, in general, the structure of companies involved is organized hierarchically back to one holding company. This company is itself often dominated by one entrepreneur.

More importantly, vertical integration also describes the way that businesses are organized. So far as the media are concerned, it refers to **the way in which functions such as the source of product, the production process, the distribution of product and the sales of product are all concentrated in the hands of one organization**. For example, Paramount can put together the package that makes up a film, including buying the rights to a book on which it is based. Then it gets the film made (though it does not own the actual studio), and it acts as distributor. Viacom which owns Paramount, also owns Blockbuster Video and has a share in the well-established US satellite movie channel HBO. Murdoch's Twentieth Century Fox is a comparable example. In this case he also owns the Metromedia chain of TV stations, and he owns Fox television which he has more or less established as a fourth major network in the States. So Fox is also into the integration of production, distribution and exhibition, typically blurring the lines between the visual media. In the print industry, Reed International is into forestry, paper production, and then the production of books and magazines from these raw materials.

This characteristic of media company ownership reinforces the point that **a lot of power is concentrated in few hands**, and raises questions about how answerable these organizations are to their consumers and to the public at large.

1.4 Conglomerates

Many of these institutions are conglomerates. This means that they are **part of a collection of companies rolled up together, not necessarily all in the media business**. Examples of conglomerates are shown in Figure 4.1. Once more we see a concentration of power, both in a given media industry and across industries. So it is that the News Corporation has interests in every sector of British media as well as in other companies. So it is that Pearsons owns the *Financial Times*, a portfolio of regional papers such as the *Oxford Mail*

SELECTED BOOK PUBLISHERS

News Internat.	Holtsbrinck	Pearson	Thompson	Reed Internat.	Random House
Harper Collins Unwin Hyman	Pan, Picador Sidgwick & Jackson Macmillan	Longman Penguin Signet	Routledge Nelson	Octopus Heineman Methuen Hamlyn	Jonathan Cape Arrow Chatto & Windus

SELECTED NATIONAL NEWSPAPERS

News Internat.	Mirror Group Newspapers	United Newspapers	Associated Newspapers	Hollinger	Guardian & Manchester Evening News
The Sun, *The Times, The Sunday Times* *The News of the World*	*The Mirror* *People* *Sunday Mirror*	*The Express* *Daily Star* *Sunday Express*	*Daily Mail* *Mail on Sunday*	*The Daily & Sunday Telegraph*	*The Guardian* *The Observer*

HOLLYWOOD FILM MAJORS

News Internat.	Sony	K. Kerkorian	Matsushita	Turner Comms Warner/Time-Life	Viacom
Fox	Columbia	MGM United Artists	MCA Universal	Warner	Paramount

Fig 4.1 Concentration of power – media ownership

through its Westminster Press, 14 per cent of Sky television and book publishers such as Longman and Penguin.

1.5 Diversification

Media institutions are also characterized by a tendency to diversify into other businesses, perhaps in order to spread their risks. And other businesses will buy into the media for the same reason.

1.6 Multinationals

These institutions are multinationals. This means that **they cross the boundaries of countries and continents**. Sometimes there are simply tax advantages in this. But it is also a way to extend power and profit. Silvio Berlusconi owns major television channels in France and Italy. If he has any kind of labour, production, or finance problem in one country then he has profits from his other enterprises to fall back on.

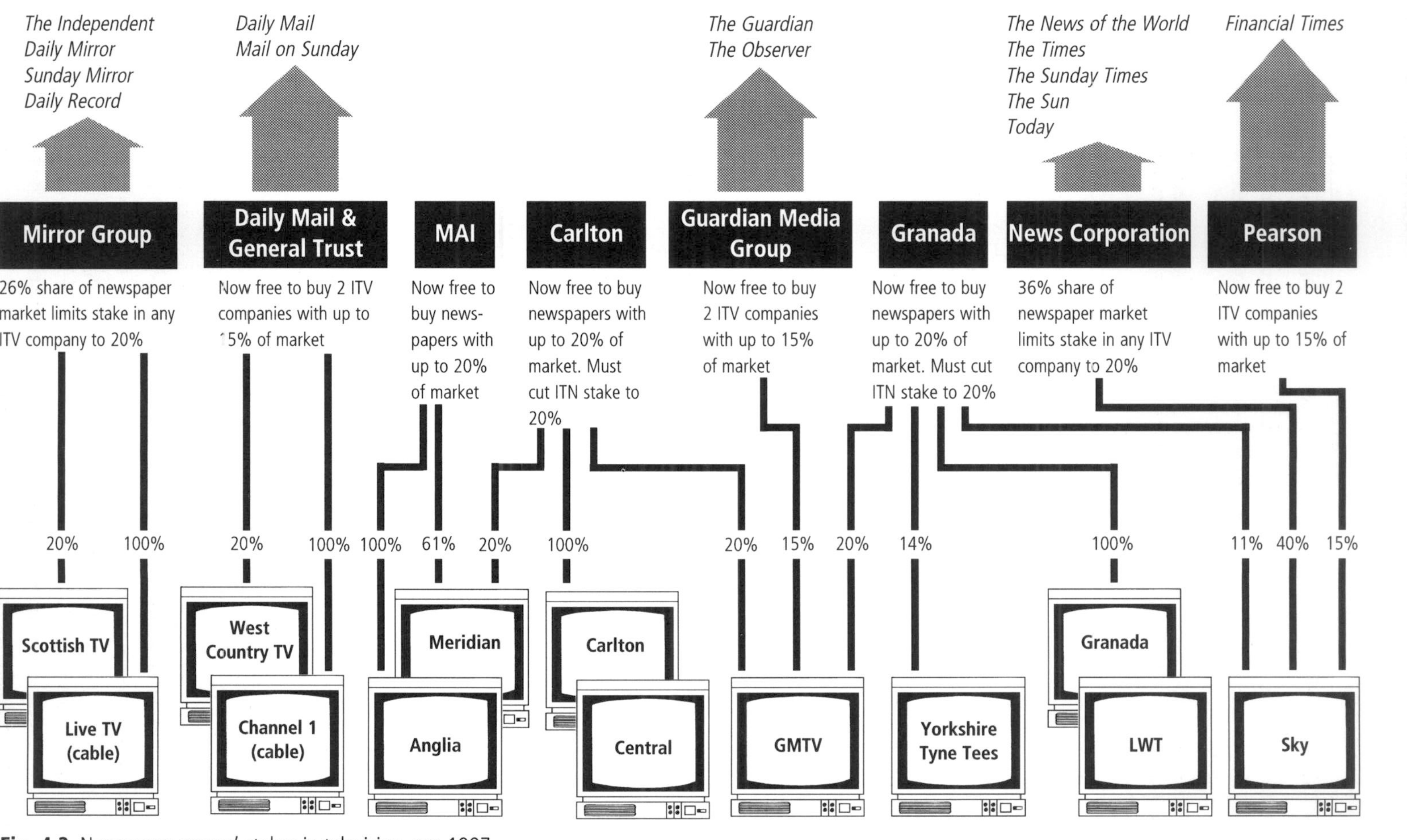

Fig. 4.2 Newspaper groups' stakes in television, pre-1997

Figure 4.2 illustrates cross-media ownership in Britain. It is significant in light of the fact that rules about the ownership of one media industry institution by another have been relaxed. They will apply from the beginning of 1997. For example, newspaper groups having less than 20 per cent of the UK market are allowed to own up to two television companies. Conversely, TV broadcasters who have less than 15 per cent of their market may buy a newspaper group (as long as it does not have more than 20 per cent of the newspaper market).

1.7 Cooperation

Even where media businesses are separate they will still cooperate in order to maximize profits. To give an example, Mills and Boon publishers, Yorkshire Television, Atlantic Video Ventures and CIC video worked together to produce a multi-media package of romance stories which were issued at the same time on video, audio tapes, books and records. It may be worth pointing out that CIC is itself owned by MCA/Universal Pictures (which is now owned by a Japanese corporation) and by Paramount Pictures.

1.8 Control and Domination

These institutions attempt to control the source of product, the means of production, the means of distribution and the outlets for their products wherever possible, because this makes their position more secure. The American film majors have started buying back into cinema chains, having been forced to sell them off 40 years ago because such all-embracing control was thought not to be in the public interest. The American majors dominate distribution and finance. Through their ability to finance films they strongly influence the kinds of films that are made, even though they do not make them all themselves. In Britain we have no film production on any scale. But there is still a powerful control over what we see because companies such as Rank, dominate distribution in collaboration with the American majors. (These majors include all the well-known Hollywood film names such as Columbia and Warner Brothers). All these British and American majors have international distribution systems. So they have a dominant position with regard to the four items named at the beginning of this paragraph. They have a dominant position in terms of deciding what we see when and where. They have a secure position with regard to profitability. I am not suggesting that such organizations abuse their power, I am simply pointing out that the way they are set up and run leads to the potential for abuse.

1.9 Institutional Values

What things do these media organizations appear to value? Certainly the companies concerned would argue that they value things like the goodwill of their consumers and the quality of their products. The problem here is that the notion of quality is arguable, and the extent of their accountability with regard to this and with regard to goodwill is also open to question. There is every evidence, as I will explain in later pages, that they have to be constrained or pressured into satisfying the full range of audience needs and maintaining product quality, rather than providing these things voluntarily.

They value **profits**. Given that most of the institutions we are talking about are funded commercially and are answerable to shareholders, then it is inevitable that the pursuit of profit is a priority.

This also means that they value **advertising**. All media industries, apart from the BBC (and perhaps not for much longer), depend on advertising to some extent to stay in business. For daily newspapers, between 30 and 60 per cent of their income comes from advertising (less for the popular press and more for the qualities). In the case of the ITV companies about 95 per cent of their income comes from advertising. It follows that they must please their advertisers, whether or not we think the effects are desirable. And it is not only direct advertising which is relevant here. Indirect selling of products through product placement is very big business in American film and television in particular. Companies pay good money to have their products used and placed in shot. So storytelling in drama, for instance, is to some extent shaped by advertisers' needs, not creative interests. The most obvious examples of this are the contrived dramatic peaks in the stories which mark the advertising break and which should hold the audience through it. Similarly, it may be that programmes or magazines will feature certain kinds of content and treatment (or leave things out) in order to please the advertisers. For example, there are fashion features in newspaper magazine supplements, or items on restaurants or entertainment in local papers, which are locked in with advertisements.

They value **audience spending power**. This means that either they will pursue large audiences – popular television quiz shows – or they will pursue upmarket wealthy audiences – quality Sunday newspapers. These newspapers may have a smaller circulation than the tabloids, but one can charge relatively more for the advertising space.

They value a **social system and ideology which is fundamentally capitalist**. This makes it likely that they will present such a system and its values in a favourable light. It is in their interests to maintain such a system because it is the one which favours the pursuit of profit and the raising of finance through shares (among other capitalist features). So the media owners tend to favour Conservative politics. Many British newspapers can be shown to favour this political party and/or its politics to some degree.

2 SOURCES OF FINANCE/COSTS

It is useful to have some idea of where the money comes from that pays for the media institutions and their products. This is because, as we saw with regard to advertising, it is at least partly true that he who pays the piper calls the tune. The amount of money available to make a TV programme or a film is going to affect what appears on screen. In other words, **the budget affects production values**. (Production values refer to the apparent quality of the product in terms of how much money appears to have been spent on what is seen on screen.) Financiers and audience may be interested in how much of the money is seen on the screen.

The topic of budgets – the proportion of money spent on different items – is a considerable one on its own. Figure 4.3 gives you some idea of amounts spent on a feature film and a drama production. Not surprisingly, budgets are a crucial consideration in the production of any media material. It is also worth

Description	Total
Story & Other Rights	$972,500
Writing	$358,500
Producer & Staff	$1,141,500
Director & Staff	$2,581,000
Talent	$3,721,000
Above-The-Line Tax & Fringe	$108,000
Total Above-The-Line	$8,881,500
Production Staff	$408,000
Camera	$385,000
Art Dept	$433,000
Set Construction-Strike	$1,235,000
Special Effects	$91,000
Set Operations	$170,500
Electrical	$393,500
Set Dressing	$634,500
Props	$81,500
Animals & Pict Vehicle	$96,500
Special Photography	$60,500
Feature – Extra Talent	$406,500
Wardrobe	$425,500
Makeup & Hairdressing	$121,500
Sound (Production)	$94,000
Locations	$944,500
Transportation	$565,500
Film – Production	$250,500
Tests	$18,000
Facilities	$386,000
Second Unit	$78,500
Special Unit	$73,500
Below-The-Line Fringe	$317,500
Total Production	$7,670,500
Editing	$490,000
Music	$670,000
Sound (Post-Production)	$275,500
Film & Stock Shots	$253,000
Titles & Opticals	$91,500
Post-Prod Fringe	$39,000
Total Post-Production	$1,819,000
Insurance	$149,500
Publicity	$51.500
Product Placement	0
General Expense	$2,090,000
Other Fringe	$3,000
Completion Bond Fee	$576,000
Deposit Interest – Income	0
Contingency	$1,876,500
Total Other	$4,796,500
Total Above-The-Line	$8,881,500
Total Below-The-Line	$14,286,000
Above & Below-The-Line	$23,167,500
Grand Total	$23,167,500

Fig. 4.3 Example of a film budget breakdown; 36 days on location, 21 days studio, 18 weeks production, rate of exchange $1.58

This budget is for a British film produced in 1993. 'Above-the-line' refers to known costs – the wages of the director, actors and crew. 'Below-the-line' refers to less predictable and controllable costs – as is evident from the list. This budget does not of course include distribution/marketing costs, which for the average Hollywood film can be 50 per cent of the production cost figure added on top.

remembering that budgets cover not only the cost of actual production, but also items such as the marketing of the magazine or film.

Let us now return to sources of finance for the various media.

2.1 Television

ITV or commercial television and radio makes most of its money from selling advertising time. In 1994 this was worth just under 2 billion pounds. Advertising rates vary enormously according to time of day and the size of the audience reached. The rates are measured in terms of TVRs or television ratings (see page 188). All rates are negotiable, even though the companies all have rate cards. But one could say that, for example, 30 seconds in peak time (say, the *News at Ten* break) nationally networked over the whole country could cost £120,000 for that one play. Alternatively, a local slot of 30 seconds for one of the smaller companies could cost only £5,000.

Commercial television also makes money from the sale of **publications** such as *TVTimes*, from **spin-off products** such as records of theme music, books and toys. About 5 per cent of its income comes from these sources. Some ITV companies finance productions **through co-production deals** with foreign networks with film majors.

BBC television also gets finance from co-productions, publications and spin-offs. But its **main source of income is from the licence fee**, which is set by Parliament, but collected 'independently' by a special branch of the Post Office Services. This licence fee in theory makes the BBC more independent of commercial pressures. But it also needs to prove itself in the battle of the ratings, to prove that it is giving the public value for money, and so help its case when asking for increases in this licence fee.

	Cost in £
Artists	36,000
Copyright	8,000
Travel	8,000
Facilities	2,000
Production salaries	32,000
Studio	24,000
Film	16,000
Design	24,000
Sets	38,000
Recording	6,000
Materials, titles, script, miscellaneous	6,000

This gives a modest cost of £200,000 for about an hour of medium-budget drama. Quality drama costs could double these figures.

Fig. 4.4 Example of a TV drama budget breakdown

The total income of the BBC in 1994/95 was £2,108.5 million, of which £305 million came from BBC Enterprises including programme sales, spin-offs and its share in UK Gold, and from which £353.4 million was spent on radio, including local radio. These figures show that ITV companies enjoy a larger income than the BBC, even allowing for a special tax called 'the Levy' which they have to pay. As a consequence ITV companies are able to spend more on programmes than the BBC.

These sums may seem astronomical, but one should remember that quality drama costs £250,000 an hour upwards to produce. Other typical costs per hour are £180,000 for light entertainment, between £60,000 and £200,000 for documentary and £30,000 for something like an afternoon quiz programme.

It is not hard to see that there are great pressures to make cheap television. Co-production is one answer for major drama and documentary series. But this produces other pressures – for example to include American stars in drama material. The usual economy is to offer repeats (which are still not cheap because of the repeat fees that musicians and actors are entitled to), or to buy in foreign material. This is often American, and may cost £30,000 an hour or even less.

2.2 Newspapers and Magazines

Their income comes from the cover price and from **advertising** space. Comment on the potential influence of advertisers has already been made. The weekly gross income for *The Sun* would be something like £5.5 million or for *The Sunday Times* about £2 million.

The gross income figures include cover and advertising. The full-page costs

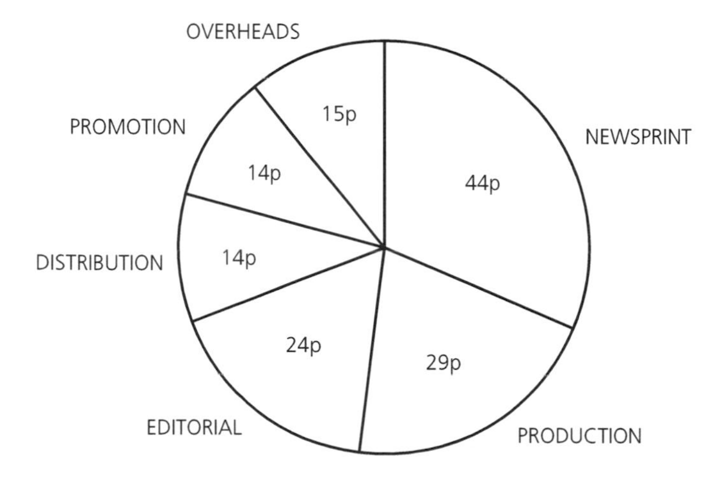

Of the £1.00 cover price, the wholesaler takes 13.5p, the retailer 33p, the publisher 53.5p. This means that there is a shortfall in costs of 40p to be made up by advertising on every copy, as well as profits to be achieved, also through advertising. This profit may be up to 50p a copy.

Fig 4.5 Breakdown of cost elements for *The Sunday Times*: each copy costs £1.40 to produce are just that – what the advertiser pays added to what the reader pays. Rates for advertisements vary. Here are some examples of full-page advertising costs: £55,000 for the *Sunday Express*, £48,000 for the *Sunday Times*, £30,000 for *The Sun*, £8,000 for a colour page in *The Sunday Times Magazine*, £5,500 for a colour page in *Elle*.

2.3 Film

The income of a typical Hollywood feature (story) film would come from the **box office returns**, from **spin-offs** such as the soundtrack music or books of the film, from television/cable rights (which may be pre-sold before the film is released), from video cassettes of the film.

The average film from a Hollywood major would now cost $30 million (£18 million), plus about $8 million (£5 million) in distribution costs.

2.4 Books

Here, much of the income comes from the cover price. A typical breakdown of expenditure would look like Figure 4.6. Income also comes from co-editions with foreign publishers, serializations and dramatizations in other media, and a variety of other sources.

2.5 Records

Their income is once more all from sales, whether single, vinyl, CD or tape – though it is worth pointing out that a well-promoted band can be earning from spin-offs such as posters and sweatshirts.

A typical breakdown of expenditure would look like Figure 4.7.

Note: it should be understood that all of these examples are indeed treated as product to be merchandised to a greater or lesser degree, and that often there is a crossover from one media to another. It is common for books to be

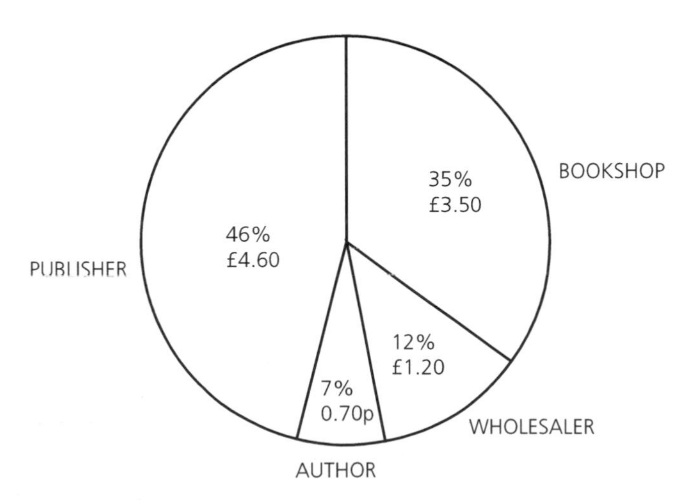

Fig. 4.6 Breakdown of average book costs by per cent and at an assumed price of £10.00

Dealer's profit	1.59
Dealer's discount	1.10
Artist's royalties	1.07
Mechanical royalties	0.52
Design and packaging	0.50
Manufacture	0.60
Arrangement and recording	0.48
Advertising and promotion	0.43
Overheads	0.83
Profit	0.30
VAT	1.57

Fig. 4.7 Breakdown of cost elements for a cassette costing £8.99

marketed as a tie-in with a popular TV drama series, or for recordings of soundtrack music to be promoted with the release of a film. At the end of all these facts and figures this should remind you that we are still talking about the kinds of institution that create the product that carries the messages. It is characteristic of them that they are indeed in the business of selling products, though this is not to say that they are unable to produce quality (or even that elusive thing called Art).

3 PATTERNS OF PRODUCTION

3.1 Routines

Whether we are talking about the production of news or of entertainment, the media organizations tend to create **routine or habitual ways of making this material**. American television, as typified by Universal Studios, has some formidable production routines, in which for most serials or series they must shoot three pages of script a day, in which everyone knows their place and their job – something like a mini production line. Even news, which has to cope with unexpected events, has routines that take it through its day (see Figure 4.8, for ITN's daily schedule). These routines or habits are useful. They provide a firm framework for coping with the technology, the time constraints, and the numbers of people involved. But they can also create habits of mind in which things are done in a certain way because they have always been done in that way.

Routines are attractive to the organizations because they make the work of production easier. Genre material (see Chapter 5) is a prime example of routine. A successful comedy series creates its own routine for making and striking sets, for assigning tasks to the crew, let alone for treatment of character. *Brookside* has created a kind of routine from its inception as a soap, with a specially built real set and a regular pattern of work for all concerned.

7.30	The Duty Editors start work. They use early radio news and newspapers in order to draft a schedule (the agenda), also based on the previous day's major stories.
	Sampling of agency material continues, as well as the daily exchange of news with other European TV news operations.
10.30	The Morning News Conference: with the duty editor, producers, other editors covering areas such as assignments and foreign news, in order to firm up the stories for the day.
11.30	The Early Bulletins Conference: with the producer, chief sub-editor, copy taster, director, newscaster, in order to make final commitments to items.
2.30	A Bulletin.
	Review of material and updates from regular sources, including reporting teams, continues.
3.25	A Bulletin.
3.45	The Look Ahead Meeting, in order to predict next day's news, to be picked up first thing next morning.
	Continue review, updates, preparation of materials.
5.45	A Bulletin.
7.00	*News at Ten* Conference, discussion of main items and presentation of programme, including newscasters.
8.00	Ideal running order is ready.
9.30	Technical run-through.
10.00	*News at Ten* broadcast.

Fig. 4.8 The daily routine for ITN News: *News at Ten*

3.2 Deadlines

All media production is always working to deadlines. In television, schedules are set up to create the material which is booked into the programme slots. These deadlines must be met or there is nothing on air. Newspapers have to be 'put to bed' by a certain time or they will not be printed in time to be put on trains or collected by vans and lorries to get to the shops in time for the morning customers. This awareness of deadlines is another reason for developing routines. The publicity machines of media organizations will be geared to the schedules which formalize the deadlines, and this publicity puts further pressure on the producers to create the material that has been promised. Magazines will have copy dates by which their material must be received and then prepared for printing. The need to meet deadlines may result in the omission or simplification of material.

3.3 Slots

Print media have space slots; broadcast media have time slots. The material available must be cut or stretched to fill these spaces. Again, this could cause a kind of distortion of the message and its meaning. Television material is often geared to the 50 minute slot or multiples of this, because the remaining time in an hour is needed for programme previews and for advertisements, especially if the programme is to be sold abroad. Press material is cut to fill column centimetres, to fit around newsworthy photographs. Items are dropped from news programmes if they fall too far down the news editor's running order and there is some major event to be covered. Short articles may appear in newspapers mainly because they will fill an awkward space on a page. Indeed any subeditor needs a supply of these fillers to be sure that the page can be made up easily. The consequence of all this is, for example, that the background to a news item may be skimped. Again, when cost is also taken into consideration, one finds that night-time television fills its slots with old films or possibly cheap chat shows. This may not be the kind of TV that the audience wants. It may or may not have much merit, but it fills the slot (and keeps costs down at a time when there are not enough of the paying audience to justify high advertising rates). You never find blank sections in magazines because the editor feels there is nothing of value to put in!

3.4 Specialized Roles

All media production is characterized by the specialized roles of the production team members. To some extent this sense of specialization is reinforced by the protective nature of the Unions, to which technical staff in particular belong. Examples of such specialization include Boom Mike Operator or Lighting Console Operator in television. Given the complexity of the task of putting together magazines or television it is not surprising that there is such specialization. But the fact is that it does have consequences for the media and their messages. One consequence is that **routine habits of production will be reinforced.**

However, these specialists still have to work together. **Media communication** is distinctive because by and large it is **produced collaboratively** and as such represents something of a consensus or a compromise view. The messages that we receive are not the creation of an individual. If there are value messages then it is the production team that actually creates them. But what they create also stands for and incorporates the values of the institution as a whole. **The team is the immediate source of the message, but they themselves are also part of the larger organization.**

3.5 New Technology

Media production is increasingly characterized by the use of new technology. Newspapers are composed electronically. They can be composed in one place, and then printed hundreds of miles away after electronic versions of the made-up pages have been sent along wires. Television material is edited electronically, it can be passed around satellites, it can mix together live action and studio material. In such ways, new technology has transformed the process through which the communication is created and distributed to the audience. **Patterns of production now depend on this technology**, most obviously in electronic news gathering and instant editing (two minutes before going on air if necessary). In some ways this new technology has opened up communication – instant electronic captions on television, for instance. In other ways it has created a new tyranny because once invested in it has to be used. For instance, if hundreds of thousands of pounds are invested in equipment and staff in order to run O/B (Outside Broadcasting), then the pressure is on to produce O/B material regardless. What's more, the way any material is handled will tend to be the way that the workforce is used to.

3.6 Marketing Product

Part of the pattern of production in the media now is a sense that one is creating product as much as communication, or a programme. The programme, newspaper or magazine is a commodity. It will only be put into production if it is seen as marketable. The irony is that a fair proportion of this marketing will take place through the media anyway. It is now common to see newspapers advertising on television to boost sales.

4 CONSEQUENCES

Having described the nature of the media institutions and the patterns of their production, it is reasonable to ask, so what? What are the consequences of elements such as concentrated ownership and repetitive production patterns?

4.1 Mass Product

One obvious result is that the producers usually tend to try to create generally successful products for large audiences. Given the size of their investment, let alone habitual attitudes favouring mass production, it is likely that they will make every effort to get their money back. Film producers will prefer to appeal to a large and international audience, especially if they spend huge sums on making the movie. One exceptional example is Kevin Costner's *Waterworld*, which is reckoned to have cost at least $200,000,000.

4.2 Targeted Product

Another consequence of the type of ownership that we have is that the **products of the various media are targeted on audiences**, indeed that audiences are themselves identified, even created (see Chapter 8). This means that films aren't made (by and large), unless the backers can recognize an audience or market for them. It means that the media makers are thinking like advertisers, trying to be specific about the type of person who is likely to want to buy their car or washing powder. It also means that media producers are less inclined to take risks, to produce material that may be excellent but which doesn't seem to fit any audience profile. They are generally happier with a film like *Interview* *with a Vampire* for a youth/horror market than with *Restoration* for a not very clear market for this British historical movie.

4.3 Repetition of Product Types

It is important, if fairly obvious, to say that material based on **a popular formula is likely to be repeated.** *EastEnders* is soap opera based on the same essential elements that make the older *Coronation Street* a hit. Scratch recording (records mixed from new material and bits of other records) emerges from amateur street culture and is packaged into new kinds of dance music whose electronic formula and semi-rap lyrics are repeated again and again by record companies for profitable consumption by a larger youth audience. Jungle music has been at least partly snatched from its independent roots by companies, and packaged for profit.

4.4 Elimination of Audiences

This is a consequence of the profit motive and the high costs of some media production. What it means is that, **unless the audience can pay enough (one way or another) for a given example of product then it will be cut or not made.** The question is, what is enough? It is a notorious fact that 25 years ago a national daily newspaper called the *Daily Herald* went bust because it was aimed at a relatively poor working-class audience and advertisers did not think it worth paying much to advertise in its pages. This meant that the cover price would have to go up. But that same audience would not have paid something like 40 pence in today's prices to buy the paper. That audience was what one might call economically insignificant. Similar audiences might be local communities who would like some of their own television. But television is expensive to operate. And so in the absence of any special subsidies through one form or another this relatively small audience receives no television product. This kind of local television was actually tried in Britain in a few towns as an experiment. But when it was seen that it would not make money, the plugs were pulled on these schemes.

On the one hand, given the fact of commercial funding for most of the media, we cannot realistically expect them to provide for every specialist audience and need. On the other hand, we must also look hard at what is called a reasonable profit. For instance, at the time of writing, CDs cost at least £10.00 each. There are many in the industry who are on record as saying that the price could easily be dropped to £8.00, and profits maintained. They would argue that the profit motive has been taken too far – in effect some of the audience are being denied the product that they would like.

So **the way money is raised to pay for media, and the way media industries are regulated in terms of how much money they may make, are crucial in deciding who gets to see and to read what.**

4.5 Exclusion of Competition

The sheer scale of most media operations, the scale of investment, and the cost of technology as well as that of production and distribution, means that to a great extent competition is excluded. If we are talking about starting a national magazine or newspaper, let alone something like satellite television, then the investment and marketing costs are huge. Clearly it is enormously expensive to join the big-time club of media producers. These kinds of cost exclude most competition and the only people who can afford to join the club are those who are already in the game. It costs something like £400 million to set up a satellite television channel. It would cost £15 million to start up a national daily newspaper but, ironically, a thousand times less to start up a small book publishing house.

Even on a more regional or local level, the costs of setting up a media operation of any kind, and running it, can be considerable. It is ironic that at the same time as the first community radio contracts were being issued (and it cost £300,000 to get London's Asian *Sunrise Radio* on the air) the more established local ILR stations were already going down the conglomerate road. For instance, London's LBC is owned by Crown Communications which itself is valued at £59 million, and as such is bigger than some regional television stations.

4.6 Polarization of Audiences

In fact, what has happened in the twentieth century, as media ownership has acquired the characteristics that I have described, is that audiences have polarized. Essentially, either **they are small specialist and relatively wealthy, or else they are very large – mass audiences**. In the magazine trade for example, you either see fashion magazines with a relatively high cover price, such as *Vogue*, or else mass-circulation magazines such as *Woman's Own*. The same is true of national daily newspapers. There is room for a new upmarket paper like *The Independent*, or for expansion of the popular press tabloids. But the middle ground has shrunk, with papers like *The Express* losing readers, and newcomers such as *Today* failing to make it.

4.7 Reduction of Choice

All of this adds up to a certain reduction of choice for the consumer. We should not exaggerate this, because we do have many specialist magazines, we do have more national daily newspapers than any other country, and we do have more and more programmes available to us, as local radio expands and satellite television comes in. But, if you consider the type of material that is presented, then it is evident that **the phrase 'more of the same' has a lot of truth in it**.

The typical pattern of ownership and control, of funding for the media, means that in cinema, for instance, we see, largely, what two industry leaders decide we are going to see. Most towns have the same films showing at any one time. And because the big distributors operate a system of 'barring' the release of films to independent operators for a long time (sometimes forever), we may wait a long time to see them at a cinema near us. For the same general reason of control the smaller audiences for more unusual films – even fairly commercial ones – may not get a chance to see these films at a cinema, or on television or video. Some films do not even leave the shelves of the studios, sometimes because a boardroom takeover means that the new studio executives don't want to release material created by their predecessors.

5 MEDIA POWER

A general consequence of the way that the media institutions operate, and of their huge financial base (and profits), is that they have a lot of power. McQuail (1994) interprets the phrase 'media power' as 'a potential for the future or a statement of probability about effects, under given conditions'. In other words, having power is having the capacity to do something, but not necessarily doing it. So it is true that **we need to distinguish between what it is theorized that the media could do and what they actually do**. This also ties in with the fact that I lead up to effects in this book, because it is only then that one can see power in evidence. Equally, we have to be careful media analysts and researchers when weighing up the evidence. It is one thing to talk largely about the power of the media, as politicians love to do. It is another thing to define and measure that power, and to be sure it is there because you can see it working, affecting people and the world.

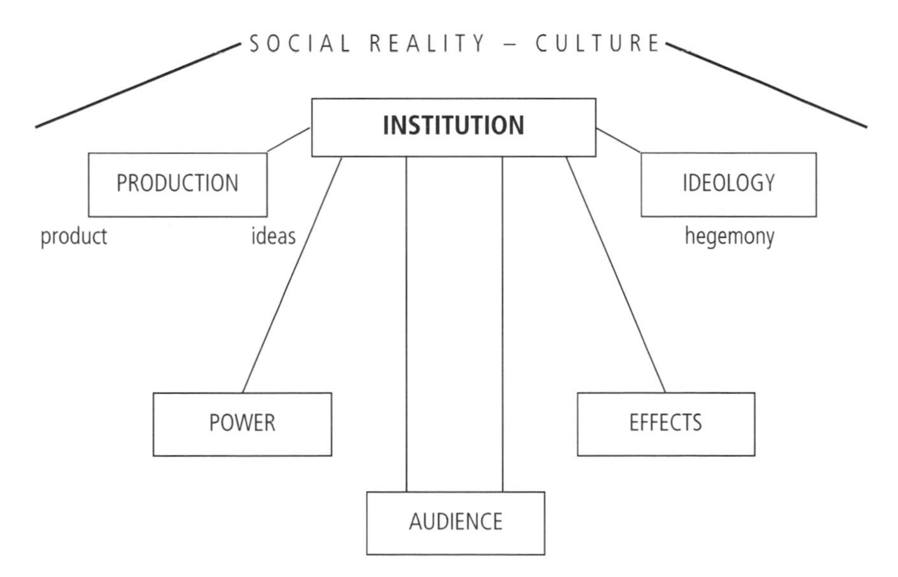

Fig. 4.9 Key concepts – Institution

5.1 Power of Monopoly

Relatively few institutions control much of the output in the major media industries. In general this means that the top five companies in each case control 50 to 90 per cent of circulation, of viewing figures, of sales in these industries, of market share. They have the power to exclude all but the richest competitors. They have the power to embark on yet more takeovers of other media industries. They have the power to resist control of their operations. They have, in the end, **the power to shape product and so to frame our view of the world.** In effect they have power over the production of ideas. Marxists would say that they control cultural production. This argument suggests that they shape what our culture is, how we see it, what ideas inform it.

5.2 Power of Owners

Certainly we should be under no illusion about the reality of this power. Newspaper owners pressure their editors, themselves people of considerable authority within the industry, to produce newspapers in a mould which they approve of. Several years ago, Harold Evans was forced to resign from *The Times* by Rupert Murdoch in a dispute over broad political policy. Conrad Black, who owns *The Daily Telegraph*, makes no bones about his power and will comment on things that he disagrees with. He said, 'What is the point in running a newspaper if you have absolutely no say?'

A contemporary European example of power and ownership is Silvio Berlusconi of Italy. He has considerable holdings in film, television and the press. He was able to use this to his advantage when conducting the political campaign that led to his election as Prime Minister. In France Robert Hersant owns *France-Soir* and the *Figaro* group of newspapers, among others, and has some interests in broadcasting. His newspapers certainly have represented political views which match his known affiliations with politicians.

5.3 Levers of Power

Marxist theory has been active in defining the meaning and application of ideology within the media. For example, the phrase **ideological state apparatus** is used to describe social institutions such as education and the media, and to suggest that such institutions work with the effect of reproducing an ideology centring on capitalism and its essential rightness.

Media power has also been seen in terms of two kinds of control.

Allocative Control is a general kind of control of the given operation in terms of allocation and use of resources, not least money. This is the kind of control that is about policy decisions. It is the control that owners and boards of management have – perhaps to hire and fire executives, certainly to determine budgets, and ultimately to decide to do things like sell up or close down operations.

Operational Control works at the level of production itself. It is the kind of working control which editors and producers have. Producers in television

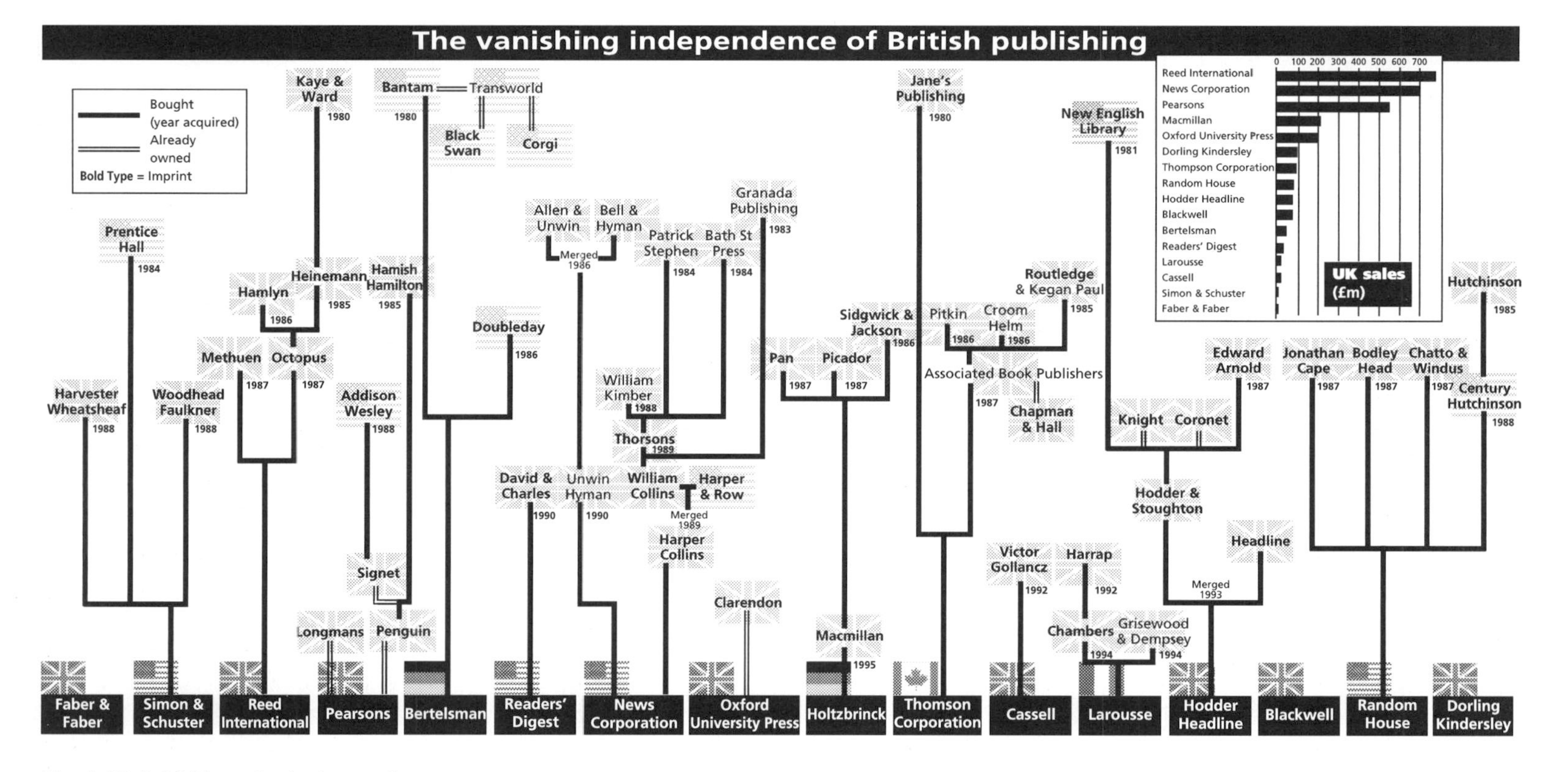

Fig. 4.10 Publishing – institutions and power

Figure 4.10 shows that British book publishing is also going through a process in which power of ownership is concentrated in fewer and fewer hands – those of multinational conglomerates.

Company	Principal activities	Employees	Finance
The News Corp. CEO Rupert Murdoch	Newspaper, magazine and book printing and publishing in Australia US, UK, Asia, China Radio, television and film operations and production (e.g. Fox and BSkyB) Information, research, data securities, couriers, data communications	approx. 30,700	Net profit 1990 A$343,305,000 Total assets A$25,000,000,000
The Thomson Corporation CEO Ken Thomson	Newspaper printing and publishing Specialised information services and professional publishing Financial services Leisure and commercial travel	approx. 45,800	Net income US$292,000,000 Total assets 1991 US$8,166,000,000
Bertelsmann AG	Over 75 book and record clubs, publisher and direct mail marketers, trade & professional markets, periodicals, encyclopædias, maps and atlases in Germany, UK, Canada, US, Switzerland, Spain, Netherlands, Italy, Belgium, New Zealand, France and Australia Music, film, television Printing and manufacturing: over 20 companies, mostly in Germany and in Europe Music and video: over 16 companies in North America, Europe, Japan Electronic Media: 23 companies, mostly in Germany Gruner & Jahr	approx. 45,100	Net income DM 510,000,000 Total assets 1990 DM 7,209,000,000
Time Warner Inc.	Publishing magazines and books Music production, distribution and copyrights Entertainment production and distribution: film and television, video cassettes, pay TV, cable and cable programming	approx. 44,000	Net income 1992 US$86,000,000 Total assets US$27,366,000,000

Source: Moody's International Manual (1993), Moody's Investment Services, New York.

Fig. 4.11 Holdings of some major media corporations

have this kind of power when they commission rewrites of scripts, or decide that some part of a programme should be dropped. To some extent this type of control is diffused. For example, a cameraperson shooting a news story or a documentary in the field won't have a producer or editor breathing down their neck. But again, they cannot stop the producer cutting their material once it gets back for editing.

5.4 Power over Product and Ideology

So **in general, media power is based on cash, legal powers and management powers.** This power then becomes the power to shape the product. The product then has the power to communicate ideology, values, ideas. These ideas then have the power to shape the views of the audience. However, at least this last stage of the power game (audience effects) is arguable. Much is believed about the power of the media over the values and attitudes of the audience. Chapters 7 and 8 talk about this, and about the difficulty of proving some kind of cause and effect.

5.5 The Power of Professionalism

One kind of power which the makers of programmes and newspapers assume is that of being 'professionals'. This is a kind of expert power which is taken on partly because of the special skills and technology which they have. But it is also elevated into something special and exclusive by these people being labelled 'professional'. The very word has come to mean something special in our society. If workers want to give themselves status, and suggest that they know and do something which no one else can do, then they like to call themselves 'professionals'. In the case of the media it has come to create something of a mystique. The word has been used for example to justify decisions made by news people about excluding information or about reporting stories in a certain way. In fact this use of professionalism is very questionable. The quality of their work, their ability to make judgements, is not evidently any better than that of many other members of our society. So this word 'professionalism' describes an assumed power which others are asked to respect, but which needs to be disputed.

Margaret Gallagher (1988), talks about three different uses of the word 'professional', all of which add up to a notion of power and authoritativeness. These are, the idea of the 'expert', the idea of the 'rational bureaucrat' (like the expert civil servant), and the idea of a special kind of worker whose business assumes 'moral values and norms' which place it and the worker apart from most other occupations.

5.6 Cultural Imperialism

This phrase refers to the way in which a culture can build empires abroad through the export of its media. The empires which are created are, it is suggested, built of ideas. When American comics, films and TV programmes are sold abroad, **the country which buys them isn't just buying stories or entertainment, it is buying the ideas or messages in that material.** These messages are to do with US values, with American ideology. This is media power over culture. It is a power which extends beyond the original audience and culture.

In his introduction to *The Media Are American*, Jeremy Tunstall said as long ago as 1977 that 'each nation at the height of its political power also had the means and the will to beam its own image around the world as Number One nation' (referring to first Britain and then the USA). He argues that without what he calls Anglo-American media domination 'many aspects of life in most of the world's countries would be different – consumption patterns, leisure, entertainment, music, the arts and literature.'

The most widely viewed pieces of media material in the world are two old American TV series, *I Love Lucy* and *Bonanza*. They have appeared in every country with a TV system, and have been much repeated. They contain ideas about family, gender, who has power in relationships. They represent American culture to these other countries. It is also the case that cheap American product can stifle the indigenous media. If it wasn't there then the media of those countries would have to develop their own product containing their own beliefs and values, not to mention their own stories.

British culture is especially prone to colonization by US media because of the language match. And so it is that our language has changed in the last 50 years to absorb American phrases. Children's games have developed out of American comic-book heroes. Some of our visual langauge is that of American cinema or advertising. The commercial power of American institutions is in our market-place, with publishers such as Random House owning British names such as Jonathan Cape.

The question of the empire of ideas is another matter. It is proposed that the British acquire US values and attitudes from the US material. But this has never really been demonstrated, however plausible it may seem. It is another area where effects are assumed but not proven. Typically, the concern is there, if only in the respect that there is a 20 per cent agreed limit on the proportion of overseas programming on television.

One should also realize that the British export their media and culture abroad, and that we acquire other cultural influences. A particular example over the last ten years has been the arrival of *Neighbours* and *Home and Away* on TV, and of *Sugar* in the teenage magazine market-place, both from Australia. British teenagers have been 'colonized' by a seductive mixture of sun, sex, romance and gossip.

6 MEDIATION

The media inevitably transform everything they deal with. Literally, they come between us the audience and the original material they use. This is a truism – a self-evident truth. What is on the screen or on the page is not the real thing, but

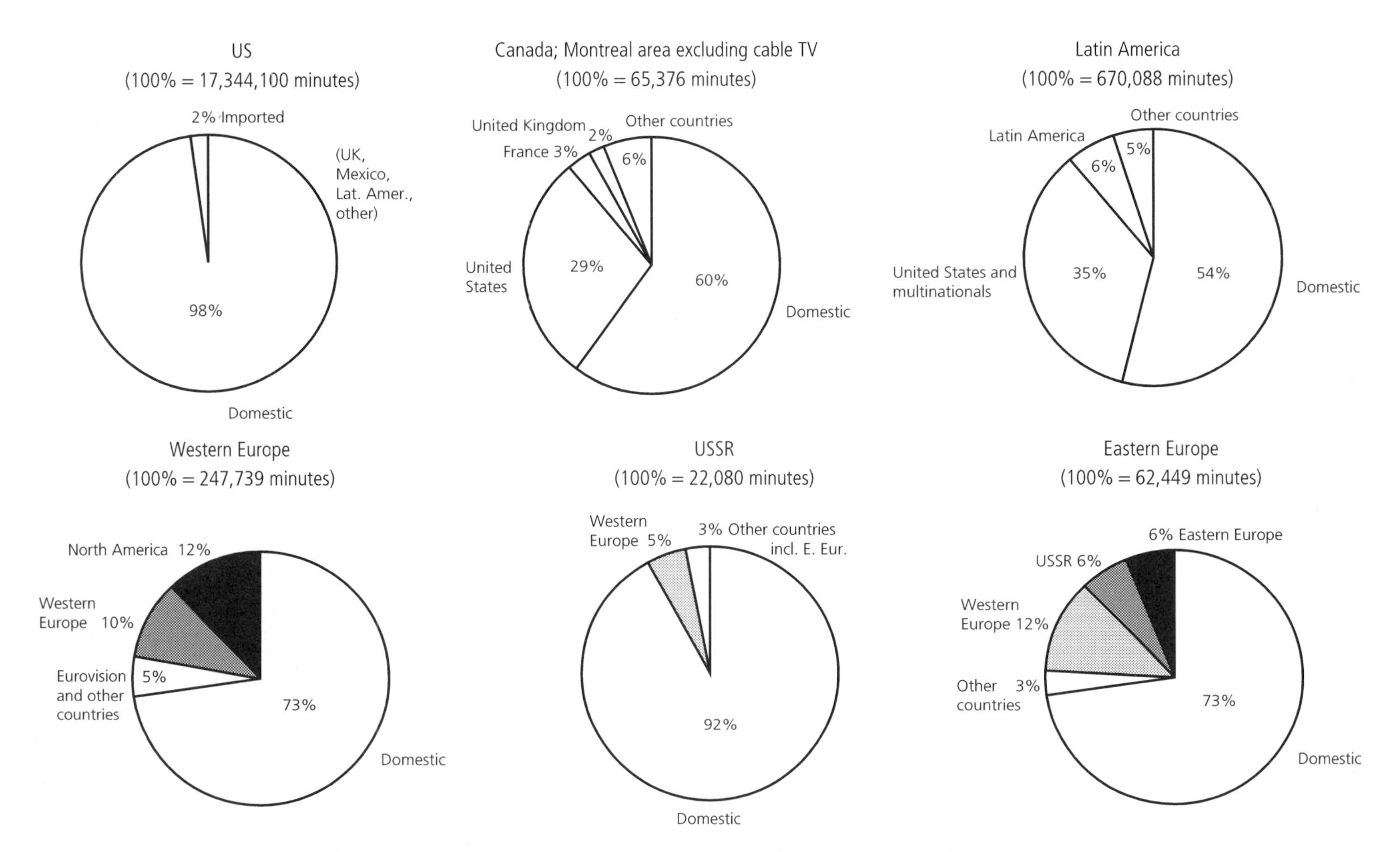

Fig. 4.12 Total programme output for selected countries/continents, in terms of sources of programme material

Figure 4.12 shows, for certain areas of the world, how much programming is indigenous and how much is imported. To an extent the pie charts suggest that things are not too bad. But then one also notes that 40 per cent of programmes in the Montreal area of Canada are not Canadian; that 40 per cent of the programmes in Latin America come from outside.

	Bulgaria		Hungary		Italy		Netherlands		Sweden	
	Min.	%	Min.	%	Min.	%	Min.	%	Min.	%
1. Drama/movies	347	90	286	79	342	90	92	88	102	94
2. Light entertainment	7	2	41	11	34	9	9	9	2	2
3. Music	18	5	5	1	4	1	0	0	0.5	0
4. Sports	0	0	23	6	0	0	0	0	0.5	0
5. News	0.5	0	2	1	0	0	0	0	0	0
6. Information	2	1	0.5	0	0	0	2	1	4	4
7. Arts/humanities/science	3	1	2	1	3	1	2	1	1	1
8. Education	6	2	1	0	0	0	0	0	0	0
9. Religious	0	0	0	0	0	0	0	0	0	0
10. Other/not attr.	0	0	0	0	0	0	0	0	0	0
TOTAL	384	100	361	100	381	100	106	100	109	100

Source: Pebren Sepstrup (1992), *Transnationalization of Television in Five European Countries*, Unesco, Paris.

Fig. 4.13 Consumption of imported programmes on national channels by main programme categories (average viewing time in minutes per week)

Figure 4.13 takes five European countries and shows how viewers there love to soak up foreign material if it is drama or films. One then needs to look at the meanings conveyed by this material, as well as to reflect on what these imports may be doing to the domestic film and television industries.

a version of it. I am not just talking about the way that events such as wars are packaged and presented for consumption in the news. The same is true of stories, for example. Sometimes this is very obvious – teledrama. The story of *Cold Comfort Farm* by Stella Gibbons is actually changed from the book to suit the demands of television. But even in the case of, say, comedy, the finished version is not the same as the original script. It is a performance which has, among other things, been directed and edited.

So the media mediate everything they touch. The fact that there is a great deal of truth in a news article about, for example, some diplomatic negotiations, does not mean that it is the whole truth. Reading the article is not the same as being there. So the only real questions to discuss are to do with the degree of mediation, how it happens, and how it may affect our understanding of the meaning of a particular example.

The idea of mediation should remind us that all media material is actually a kind of **representation.** This means that it is a **construction.** It is not the real thing. It is not the original idea, the original experience, the original object. It is artificial. It is something constructed from sets of signs. The TV image of the car, whether in a thriller or in a programme about a car show, is not the car itself. It is a representation of the car through a picture code. In the same way, a journalist's article about cars at the car show is just a representation of the cars in a written verbal code. This difference between real life and the artificial medium is vital. Especially when we are dealing with picture media, we are inclined to forget that the 'real'-looking picture is only a version of reality – something which has bias built into it. You should bear this in mind when reading the section on Realism in Chapter 5.

One question which mediation raises is that of intentionality. From one view it could be said that mediation is unavoidable, it is an inevitable aspect of the process of manufacturing media product. Other views would take on the decision-making/choices aspect of constructing product. For example, Skeggs and Mundy in *The Media* (1992) refer to critiques offered by others when they say that 'every use of the media presupposes manipulation'. In that case we must ask ourselves who does the manipulating, how and why?

Part of the answer is that those who censor material in one way or another are also manipulating. And factors which constrain media production are also agents of mediation.

7 CONSTRAINTS

7.1 Definitions

The idea of constraints relates to something which restrains or holds back. When looking at the kind of media ownership we have we then need to consider constraints as part of this. Limits on the ability of the media institutions to do what they want actually shape the kind of institutions that we are talking about. They are not the same kind of organization that they would be without such constraints.

Broadly, constraints can be described as **voluntary or imposed, and internal or external.** Voluntary internal constraints would include the codes of practice which broadcasting uses to guide its operations. Imposed external constraints would include the D-notice system through which the government tries to cut off reporting of items which it considers to be 'contrary to the public interest' (with the implied threat that the Official Secrets Act might be used against those who report or publish in the face of such a notice). In a culture which believes in freedom of speech and the rights of the individual, it is perhaps not surprising to see that there are relatively few formal and external constraints on media operations. There are other books which go into this topic in more detail, usually under the heading of 'censorship'. What this section will do is to give you knowledge of the basic types of constraint and their implications.

7.2 Self-regulation

In terms of output, the media are almost entirely self-regulating. That is to say, **the content and treatment of articles, programmes and advertisements is monitored and vetted by bodies set up by the industries themselves.**

There is –

- the **Advertising Standards Authority** for advertising (but please remember that advertising is not a form or medium of communication, only a way of using media)
- the **Independent Television Commission** for commercial television and radio
- the **Press Complaints Commission** for newspapers
- a committee system answerable to **the Governors for BBC television and radio**
- the **British Board of Film Classification** for cinema.

These bodies act as constraints because they set standards published in pamphlets and interpreted by them as they check complaints and check the product. Usually examples which the makers are doubtful about will be referred to such bodies. But it should be clear that these self-censoring organizations are not directly answerable to the government or to their consumers in any way. Their power to enforce decisions is very variable. The ITC committees which vet programme schedules in advance, review questionable programmes, and many advertisements (including those going out nationally) are ultimately backed by the power to stop transmission, because the ITC (not the TV companies) runs the transmission system. This power has never been used, but in principle the ITC could order its engineers to switch off the line to the transmitters if one company tried to send an 'unacceptable' programme from its studios. This power derives from the Act of Parliament which sets up the commercial system and which forces the commercial companies to pay for the ITC whether they like it or not. On the other hand, the Press Council is pretty toothless. It may hand down formal-sounding

	News/Factual			Fictional/Entertainment		
	ITV	C4	C/S*	ITV	C4	C/S*
Accuracy	207	35	4	49	7	1
Impartiality	44	17	2	1	1	–
Sexual Portrayal	23	33	7	165	38	8
Language	26	14	1	120	30	10
Violence	20	3	8	126	9	8
Other Taste and Decency	190	103	8	293	59	15
Other Unfairness	281	50	–	78	8	2
Racial Offence	18	11	1	18	6	1
Religious Offence	9	2	1	36	1569	–
Scheduling	121	31	3	124	20	3
Regionality	26	2	–	–	–	–
Miscellaneous	148	67	35	127	19	5
Sub-totals	1113	368	70	1137	1766	53
Total complaints = 4,507						

* Cable and Satellite
Source: ITC Annual Report & Accounts, 1995.

Fig. 4.14 Total number of complaints about TV programmes

Figure 4.14 is about the number of complaints received by the ITC about programmes broadcast through the commercial system. First, in relation to the huge number of viewers in a year, the number is tiny. Ask yourself why one might not believe that, from the fuss that has been made in the press about some programmes. Who is making the fuss? What have they to gain from making a fuss? Why might they be calling for censorship?

judgements about the propriety of certain articles, but it can do nothing to bring the newspapers into line because it is in fact set up and paid for by them.

7.3 Broadcasting Standards Council

Existing in 'shadow' since 1988, this body became statutory under the Broadcasting Act, 1990. Its role is to 'consider' the portrayal of violence, sexual conduct, and matters of taste and decency in broadcast material of all kinds. It examines complaints, publishing its findings in a monthly bulletin. Although advisory only, these findings act as a form of pressure on the industry. The Council can order broadcasters, at their own expense, to publish findings either on air or in the press. Its independent role is unusual since, with the exception of the Broadcasting Complaints Commission, the broadcast media in Britain have been self-regulated. The Council may make reports to Government on aspects of programmes falling within its remit: an example was its report in 1992 on the satellite porn channel, Red Hot Dutch. Original proposals for this body led to fears that it would be more censorious than has been the case. In 1993/94, it completed consideration of 1,174 complaints, upholding in full or in part 329 of them (28 per cent). Examples of complaints upheld in 1994 include the depiction of a suicide in a Channel 4 play shown before the 9.00 pm watershed and, in another case, the use of the 'F word' in a Radio 1 FM lunchtime show.

7.4 Editorial Control

In fact most of the real self-regulation takes place at editorial or programme level. The people who produce and make the programmes have a good idea of what is allowed and disallowed through, for example, codes of practice published by the broadcasting organizations. Nevertheless, it gives food for thought that, for instance, in spite of this, 20 to 30 per cent of the advertisements presented to the ITC committee are rejected for one reason or another. This does not necessarily mean that all those advertisements are offensive, any more than are programmes which we never get to see. You or I might actually approve of them. The fact is that you will never know or be able to judge.

Since media industries are dominantly self-regulated, editorial control of material is a very important constraint which defines what kind of papers and programmes we receive. In the case of the news arms of these industries we are talking literally about editors. But magazines and books also operate through editors, and producers can have a similar constraining effect on programmes. News editors act as a kind of filter (backed up by their sub-editors who do most of the donkey work of selecting and rewriting general news items to fit the space or slot). Producers can make decisions to change or leave out material. The BBC in particular uses a system of 'referral upwards' which means that, if in doubt, the producer asks a superior to check out material, perhaps in terms of its offending listeners' sense of good taste in the case of radio. In this way, the real constraint is the sense of values that permeates the institution – it is this sense of what will or will not be accepted by those in control at the top which shapes (and constrains) the decisions of those with direct responsibility for making programmes. In other words **institutional values act as a constraint.**

7.5 Legal Constraints

The most formal constraints on the operation of the media come through the laws which set up the relevant bodies in the cases of broadcasting and satellite transmission, and which cover specific issues such as Libel, Slander and Official Secrets. Remember that no one may broadcast anything without a licence from the government. So commercial broadcasting based on licences granted by the ITC is constrained by the terms of the Broadcasting Act of 1990. By definition, pirate radio stations are those which operate without such a licence. The Broadcasting Act (and the broadly similar Royal Charter enabling the BBC to operate) is in fact pretty general in terms of what it says about programme material, using phrases such as 'not to give offence', or 'within the bounds of good taste'. The actual interpretation of these phrases brings us back to the voluntary bodies already described. There are no comparable acts setting up or describing what the press institutions may or may not do.

The other laws referred to below mean the media organizations need legal departments to advise on what is or is not likely to give rise to a legal case. The goalposts tend to move according to judgements of the day. *Gay News* was successfully prosecuted by an MP invoking the old-fashioned notion of blasphemy as covered by the Obscene Publications Act for publishing a poem about Christ as a homosexual. But some other attempts to use this Act have failed. The Government fought a long battle in 1988/89 to stop Peter Wright's book *Spycatcher* being published anywhere in the world, because it claimed it breached the Official Secrets Act in what it revealed about Britain's spying activities in the 1960s in particular. It eventually lost the case against newspapers for publishing parts of the book. It won the case against British book publishers (or rather won an injunction forbidding publication). But it lost a famous case in Australia against other publishers. To take another example, in 1989 there were a number of successful prosecutions for libel against tabloid newspapers for publishing certain allegations about the private lives of various celebrities. The size of the damages (a quarter of a million pounds or more) suggests that this law is being interpreted more strictly than before.

Relevant acts may be briefly summarized as:

- the **Laws of Defamation of Character** (slander for the spoken word, libel for the written word);
- the **Official Secrets Act (1989):** this is usually operated through the D-notice system in which the Home Secretary's office puts such a notice on material which it does not want to be published or broadcast (see above);
- the **Young Persons Harmful Publications Act (1955):** which was directed towards comics and magazines in particular, and is intended to control the production of horrific or otherwise harmful material;
- the **Obscene Publications Act (1959);**
- the **Prevention of Terrorism Act (1974):** generally geared towards control of information about and coverage of events in Northern Ireland;
- the **Contempt of Court Act (1981):** forbidding the publication of anything relating to a trial in progress which might prejudice its outcome;
- the **Video Recordings Act (1984):** restricting what may be hired from video shops, and imposing categories on films on video.

There are also particular provisions of other acts which may in various ways constrain the news operations of the media. For example, the Criminal Justice Acts forbid publication of evidence against people sent for trial from lower courts.

7.6 Financial Constraints

It is also worth remembering that there are practical constraints on what we see or do not see, what we read or do not read. Obviously, **the amount of**

money or the amount of time available to a programme will affect what it is like. 'Market forces' are also a kind of constraint. That is to say, something which is popular and profitable is more likely to be produced than something which is not. For example, with the advent of deregulated broadcasting, and as more organizations have the freedom to make programmes, there are serious worries about who will pay for expensive television drama. Maybe we will be deprived of one-off plays by famous authors and will if anything, see more popular fare such as Australian or American mini-series.

8 CENSORSHIP

This section is brief because for the most part the relevant points have been made under the heading of Constraints. Censorship is usually thought to be about **direct removal, part removal or change of material that offends the censor.** A censor has the power to enforce his actions and decisions. **There is virtually no direct censorship of the media**. It is done for the most part through the self-regulating bodies described above. These are not directly answerable to the Government or to the public. However, it is worth remembering that there are some examples of specific external and direct censorship. These are imposed by the Home Secretary through some short Act of Parliament or an amendment to a previous act. A recent one was the requirement that broadcasters should not transmit soundtracks of utterances by members of the Irish Sinn Fein organization, and that they should not screen interviews with these people. The Government decided that such reporting provided Sinn Fein with 'the oxygen of publicity', and that this was not in the public interest. This ban, now lifted, was an example of contradictions in the dominant ideology because it ran counter to beliefs in free speech and in the public's 'right to know'.

So **censorship is mainly internal, and variable in its strength**. The British Board of Film Certification awards certificates to films, without which they cannot have normal public viewing. It also decides which age groups may watch which films. This control is quite powerful. It can stop communication before it reaches the audience. On the other hand, the Press Council has no such power. It can only express opinions about material after it has been published and in response to complaints. It has no power to force any of the newspapers to do anything. The ITC, for the commercial broadcasting system, falls somewhere between these two models. Technically it is independent because it is a body paid for by the companies (as a condition of their operations) but which is not owned by any of them. It discusses programming with the major companies, relies on them to conduct much of their own programme monitoring under ITC guidelines, but also samples materials. If the ITC governors advised by the monitoring committee decide that some programme should not go out, then they will make this request (backed by the power to withdraw an operator's licence).

One problem in making sense of censorship is its relative **inconsistency**. If we take the Official Secrets Act (see legal constraints above), then there are two famous cases in the 1980s which had very different results. In both cases civil servants were actually taken to court for leaking information. So we are not dealing with censorship as such, but only with indirect censorship – how the results of the cases might affect what the media decide to put out another time. This is why it is easier and more accurate to talk about constraints rather than censorship. In 1984 Sarah Tisdall was sent to prison for leaking information that the Minister of Defence allegedly intended to deceive Parliament about the arrival of American Cruise missiles until it was too late. In 1985 Clive Ponting was acquitted in a court case in which the Government wanted him to be put in prison for revealing that the same minister had also allegedly deceived Parliament about the circumstances of the sinking of an Argentinian battle-cruiser during the Falklands War.

So there is some kind of censorship of media communication. But it does not, as the word suggests, operate through some form of centralized vetting through which every single item has to pass in order to be approved. You should refer back to the section on Constraints to get a better picture of how communication through the media is pressured, filtered, denied and shaped in various ways.

9 MEDIA AND GOVERNMENT

It is worth looking briefly at the relationship between media and Government, not least because in general the latter acts as a constraint on the former. These two institutions exert great power in our society, and are in a sense jealous of each other's power. Government seeks to curb the freedom of the media to say and do what they like – in the name of the people whom they represent. But the media also claim to represent people, and would often like to be free of restraining power. The following summary of points of contact between these two institutions includes some of the factors discussed above.

The Law. Through invoking laws like the Official Secrets Act to protect its information and operations, various arms of Government (mainly Secret Service operations) come up against the media and its belief in its right to say what it will.

Press Conferences and Press Releases. Government departments (and Ministers) are a major source of information for media news operations. To this extent they are in the hands of the Government when it comes to getting newsworthy information via the relevant Press Officers.

Lobby Briefing. The two main parties operate this system, which many of the media news people resent. In effect the Chief Whips of the political parties have a list of acceptable journalists to whom they will give information (sometimes advance leaks) at such special briefings. Newspapers like *The Guardian* have taken to boycotting this system as far as possible. They feel that it is a misuse of Government power over sources of information, which should be freely available to everyone equally.

Appointments. It has already been said that the Government has a kind of power in its ability to make senior appointments to the broadcasting systems in this country.

Finance. It has been pointed out that the BBC at least is indirectly beholden to government because its income from the licence fee depends on the agreement of Parliament to set it and increase it.

Direct Control. The Government reserves certain powers, again in respect of broadcasting, to demand air time at the discretion of the Home Secretary.

Indirect Control. This is very difficult to prove, and usually only emerges when there is a public argument between media and Government, over reporting from Northern Ireland for example. At such times it has been revealed that conversations take place between senior officials of the Government and of the relevant broadcasting organization in which it is suggested that the programme concerned should not be broadcast. The ability of senior broadcasting officials to resist such pressure is itself a matter of concern, given the influence of politicians over appointments.

Media Appearance. Another obvious point of contact are the appearances of politicians, including members of the Government, on news and current affairs programmes. This is a good example of the mutual interests of these two powerful institutions. The broadcasters need the politicians as sources of information, as personalities to give their programmes credibility. The politicians need the appearances to give themselves credibility, in their own party as much as among the electorate. Broadcasters keep a list of such people whom they regard as sound 'performers' and who may be called upon to make interesting appearances in debates and interviews.

Party Political Broadcasts. The political parties may demand these as of right. This is some measure of their power in the continuing tension between media and Government. Traditionally the amount of air time, especially around elections, is something which is carved up between the Chief Whips of the Tories and the Labour Party and senior broadcasters. This deal has been under strain in recent years as more parties of the centre have emerged and have expressed discontent with their share in these arrangements. It illustrates how delicate is the power balance between media and Government, and how far it has depended on a consensus between the two. Some people argue that this consensus depends too much on the people concerned sharing common social and educational backgrounds.

All these points of contact raise a number of issues. These are essentially about the degree of power which politicians should have over the freedom of the media to say what they will, and the rights of the media to free speech, given the fact that what they 'say' and how they say it may influence the political views of the audience/electorate. There is concern over politicians' manipulation of their power over information through devices like the timing of press releases to coincide with news broadcasts. There is concern over the extent to which the media are representing political issues in terms of personalities rather than in terms of the real arguments involved.

10 PROPAGANDA

This word is thrown around inaccurately, not least where politics is concerned. People may talk loosely about government propaganda, when they don't like what they hear on a radio Party broadcast.

In fact **true propaganda, like true censorship, depends on centralized control of all sources of information about a topic.** So propaganda as such does not really exist in Britain.

Also, the term tends to be associated with advertising. To this extent it is defined as powerful and persuasive communication. But again it has to be said that advertising is not, in general, propaganda. You don't have to buy the car being advertised, and you can find out more about it if you really try.

One might have a case for arguing propaganda where it is not easy to find alternative views and information on a given subject. An example would be the government advertising campaign to persuade parents to have their children vaccinated against whooping cough. This campaign became a media issue as it happens because a few vaccinated children became ill, and then there was a public debate about the justification for the vaccination programme and the way in which the public had been 'persuaded' that it was necessary.

Significantly, the clearest examples of propaganda in this country have been seen in times of war. In the last world war Britain set up a ministry of information not only to censor and filter information, but also positively to produce media material which persuaded the public of all sorts of things, most obviously of the importance of aspects of the war effort, such as 'digging for victory'. Because such persuasive and manipulative uses of the media were 'on our side', the word propaganda is not always associated with this kind of communication. Propaganda is so often thought to be what the other lot do. This is not true. Indeed it is part of recognizing the existence of ideology, of our particular way of looking at the world, which helps students of the media understand what is going on 'in their own back yard'.

The Falklands War of 1983 is a more recent example of propaganda in war time, when the Government again set up a system to control, censor, suppress, manipulate information to be released to the media, in order to both support the war effort, but also to support its view of what was going on and why. To this end it did, for example, delay information about a damaging rocket attack on a British ship. It claimed that photographs could not be relayed back to Britain because of technical difficulties, which was untrue.

It is also arguable that government operates something approaching propaganda in peacetime if it uses denial of information (via press conferences, for instance) to represent selective views. This is the point at which censorship becomes part of propaganda. **What we don't know is as important as what we are told.** The fact that we rely on the media by and large for what we do know about our governments and their operations reinforces the importance of the media in our societies. For example, there was a major report and enquiry in 1996 (the Scott report) into the matter of the sale of arms to Iraq, which was forbidden. The enquiry took place as a result of media reporting and pressure following investigation of government complicity in enabling illegal sales to take place, and following a trial in which business people were acquitted of making illegal sales because it was clear that government had endorsed these (and because the trial judge overrode government claims to the right to suppress relevant documents on grounds of national security). There is a heavy fog surrounding the question of who was right to sign what paper and how far government and business were working within the law and the government's own rules. What is clear is that the British people would never have known about the matter if at least some of the media had not pursued their investigative functions in the face of political attempts at a form of propaganda.

Anything approaching a definition of propaganda which comes through news media is of course of particular concern. Even tabloid newspapers are regarded by many as offering 'truth' at least in respect of hard news stories. McQuail in *Media Performance* (1992) talks about 'the near impossibility of identifying it [propaganda] in news output in any certain or systematic way'. He refers to the fact that it may not be the channel from which the propaganda comes, but rather the source which the channel uses – e.g. the people who gave the press conference. It is true that is not easy to notice or prove a sustained campaign of deliberate misinformation from some concealed source. At the same time, when associating the idea of propaganda with bias, he does point to particular presentation devices and uses of language which signal that something is going on. It is these devices which the media student needs to be aware of, and to interpret. He refers to 'flattering language; non-attributed sources; suspicious juxtaposition of items'. So I would argue that if textual analysis throws up signs that someone wants us to read a text in a particular way, then this suggests that something like propaganda is happening. It is then another matter to determine the source and the real intentions of that source. (See also 'news bias' in a later chapter.)

11 FUNCTIONS OF THE MEDIA

11.1 General

The word 'functions' covers **what the media are there to do, what they actually do and what their purpose seems to be**. It is, for example, the function of a careers service to give advice to people. The media themselves have their own view of what their functions are. Generally of course these are seen by them in a positive light. For example the various television acts which establish commercial broadcasting refer to requirements to provide a news information service, to maintain proportions of British material, to observe impartiality, and so on. But how things work out in practice may seem rather different. So what are the functions of the media?

For a start, that rather depends on your own beliefs about the media, since their functions are not written down explicitly. It also depends on intention –

what they intend to do, and what they actually do, in your view. If we take a commercial view driven by viewing and readership figures then one might say that the media only function to satisfy our needs. But it isn't that simple. The best way to sort it out is to look at broad categories of media activity, and then at some different interpretations within these categories. So, if the media function to provide information, what is done with this information might look rather different from the points of view of a newspaper owner and a shipyard worker, from the points of view of a Conservative MP and a Labour councillor.

The following list summarizes ideas about media functions. The type of function is followed by opposing views of what that function may mean in practice. These opposing or negative views are sometimes described as **dysfunctions.** This should give you some ideas for discussion about how the media institutions may be judged in terms of what they do with their communication.

This reference to discussion points up the fact that one is also talking about what are called **media debates.** In effect a debate is a discussion about a key question relating to some aspect of the media. Examples of such questions connected with the function sections which follow would be:

- Should the media inform us as well as entertain?
- Do the media reflect our culture or do they shape it?
- Do the media give us a fair range of views across the political spectrum?

The media themselves often act as a forum for discussion of these questions. They raise debates and provide a platform, they report such debates carried on by other bodies – Parliament, for instance. In fact one could say that one of their functions is (or should be?) to raise debates about themselves.

The large institutions of media ownership do perform various functions as they publish newspapers, broadcast programmes, and so on. What they think they do, what they actually do, and what they should do, is all a matter for dicussion.

11.2 Entertainment Functions

The media provide entertainment and diversion for their audience.

- This entertainment functions to provide healthy amusement and pleasure for the audience.
- This entertainment functions to divert audience attention from serious social issues and inequalities.

11.3 Information Functions

The media provide necessary information about the world for the audience.

- This information functions to help us form a view of the world in geographical, social and political terms.
- This information functions to structure a particular view of the world and to pacify the audience.

11.4 Cultural Functions

The media provide material which reflects our culture and becomes part of it.

- This material maintains and transmits our culture; it provides continuity for that culture.
- This material develops mass culture at the expense of the diversity of subcultures.
- This material can maintain the status quo in cultural terms, but may also discourage change and growth.

11.5 Social Functions

The media provide examples of our society, of social interaction, and of social groups.

- These examples socialize us into beliefs and relationships which help us operate successfully as members of society.
- These examples socialize us into beliefs and relationships which naturalize one view of society and stop us obtaining and acting on any alternative views.
- These examples serve a kind of function called correlation, relating one event to another for us; putting together events and a sense of what society is and what it means.

11.6 Political Functions

The media provide evidence of political events, issues and activities.

- This evidence enables us to understand the operation of politics in our society and to work more constructively in that political process.
- This evidence gives us the illusion of participating in political process, but actually endorses the authority of those who continue to run our lives unquestioned.
- The media are capable of mobilizing public opinion; that is to say they can raise issues which the public may not have thought of, and they can suggest a way of looking at those issues. In this way, the media are also capable of shaping opinions about political events and issues.
- In wartime in particular (e.g. the Falklands War) the media serve a political function of propaganda, not least because the government then controls sources of information.

The nature of the institutions and their operations helps shape the media product. The proposed functions of the media may be an expression of their values and of the way they work. Media product may be an expression of those functions. For example, if a media organization sees its main function as to entertain the audience at all costs, then it is likely to try to produce a lot of fun material that will appeal to a lot of people. This seems fine in principle, but in practice it may produce a lot of 'rubbish' – you discuss what you think is garbage, and what is not!

Certainly the people who license British television organisations think this could happen. They believe that the media should carry out an information function, that they should function to please minorities as well as the majority. So they write this into the contracts.

12 MEDIA AND NEW TECHNOLOGY

Institutions of the media have been transformed since the 1980s by the advent of new technology (NT). In particular this has related to developments in and applications of computing power. It has affected how material is made, how it gets to the audience and has even created new media. It has affected aspects of media operations such as financial management which are not evident to the audience. Given that one could write another book about the subject, I want to look only at a few aspects of this change, which is of course still going on.

New media have created new institutions, expanded old ones and have created new product to sell to the audience. CDs have virtually killed off vinyl and have involved the production of a whole new technology (and market) selling the equipment to play them on. Now CD ROMs have arrived as computing power becomes the norm at home, and with the effect of displacing the book as a source of information. Nintendo and Sega have arrived as typically worldwide institutions dominating the new media of computing games machines.

The process of **media production** depends on applications of NT. This book is written and composed via computers. It can be 'typeset' in one country and printed in another. Lighting in television studios is controlled via computers. News is gathered via cable and satellite links. Film rushes can be replayed instantly from slave video cameras linked to the film cameras. We now have non-linear editing (in computers) of short film sequences such as advertisements. Even feature films are now being edited in this way, in segments, before being translated back to celluloid.

New forms of distribution are changing the way that we access media material around the world. The Internet is an unusual example of a system which can access text, pictures, pop music (but not yet films and programmes) literally on a global basis and at very little cost. It is unusual because it is not institutionalized. No one owns it. It exists in a cyberspace spun out of existing cables and satellite links and focused on the computer memories and programmes of millions of machines around the world. A more mundane example is that of satellites. The satellites are put up by giant telecommunication companies in conjunction with military authorities or governments. The channels on the satellites are owned by developing organisations where typically even the BBC goes into co-ownership to run a channel called UK Gold.

The expansion of means of distribution has enormous implications for culture and for the commissioning of product. In 1988 the government said in a white paper, 'as delivery systems proliferate, national frontiers begin to blur

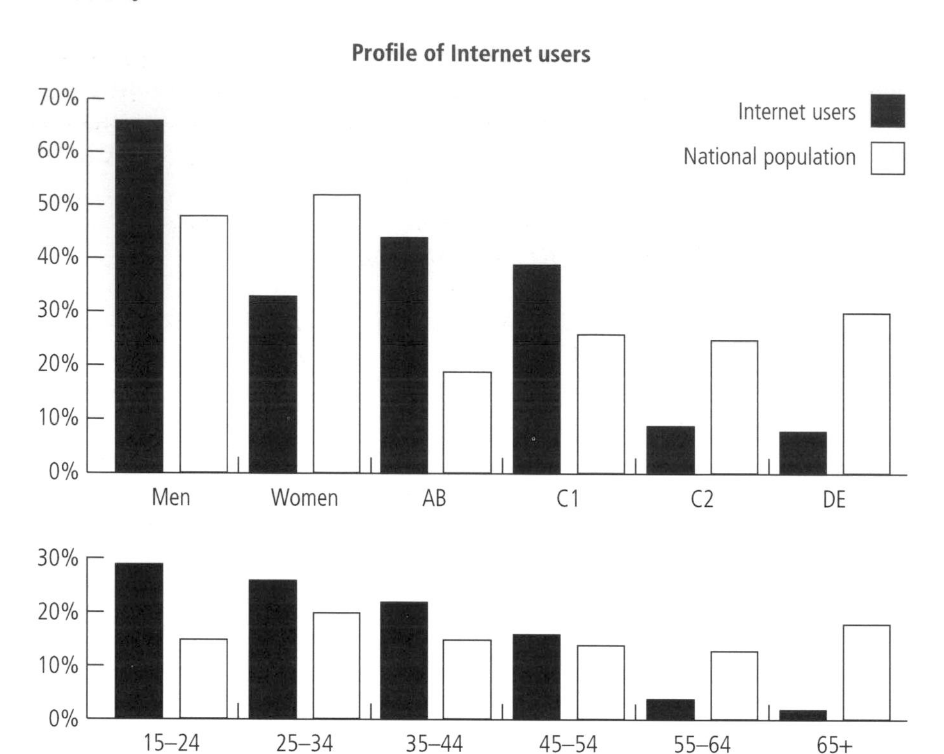

Source: Continental Research.

Fig. 4.15 Internet users – the new media

Figure 4.15 makes it pretty clear that present use of the Net is dominated by younger well off males. The researchers reckoned that about one million people have access to the Net at home. The Net is not an obvious candidate as one of the new media, but then again it can provide both entertainment and information. So ask yourself, what makes it different from television or from magazines?

or disappear. There will be increasing demand from an international market for programme material.' It is NT which provides the delivery systems. Co-productions are becoming more and more common, internationally (*Life in the Big Freezer*), and even within Britain where commercial and BBC channels collaborated to produce and screen Dennis Potter's last TV play, *Karaoke*. Ironically, a play by Potter is not the best example of cultural effects and dangers – something which Potter himself famously inveighed against. The danger is that if NT opens up international markets, then it is those markets and those audiences which the media producers may feel they have to please – the kind of problem that has already mortally wounded the British film industry, which is continually defining itself by reference to Hollywood.

Media product is transformed by NT in ways that steal up on us and which we take for granted unless we can remember a time when things were not done in a certain way. For example films such as *Toy Story* (1996) involve computer animation and computer matteing which were not possible even two years before. They create a new version of reality for the next generation of viewers, where the impossible becomes routine. It is generally agreed that it is only a matter of time before the authenticity of the computer-generated image converges with that of the reproduced image. And even reproduced images are now being captured ever more frequently by electronic 'movie' or still cameras. Such computing control of visuals has changed everything we watch. This is fairly obvious in some television title sequences. It is not so obvious in the way that photographs are routinely touched up in magazines, most obviously to bring us idealized images of people, rather than the 'real' people themselves.

Everywhere we turn to look at media products NT has changed what we see. Advertisements are full of images which mix the real world with computer-generated images. The news is full of computer-generated graphics. The weather moves across our screens in simulated images. Newspaper and magazine pages mix text and image in a layout which is courtesy of the computer.

This idea of **convergence** is possibly the most important concept of all if one is trying to make sense of NT and media. Not only are the institutions converging in the sense that they all take stakes in one another's industries, but also the technology is converging so that the idea of separate forms of communication is becoming less and less true. In particular one can see that words and pictures and music are now all available on CD ROM. And CD ROMs can be played via computers which themselves can be linked to screens and audio systems. Even films can now be compressed on to these discs. It is not hard to see the time when video tape and even possibly film as celluloid in cinemas will become redundant.

The convergence of institutions means a magnification of power over product and ideas within the product. The convergence of technology for reproducing texts could mean that a lot more is available more easily to a lot more people. However, the fact that all forms of communication are encoded in the digital language of machines also makes it easier to manipulate them, to change 'what is said'. Whatever happens with the control of information, ideas, and entertainment, clearly existing trends towards a visual culture, towards more and more kinds of entertainment, will continue.

Issues of access, choice and copyright are also raised by NT, which offers potentially a huge diversity for the consumer, but headaches in terms of protecting property and collecting money for the 'new media'. Lorimer discusses this in *Mass Communications* (1994) where he points out that the development of cable TV has been hampered to a degree by the self-interest of the established broadcast technologies. Similarly, there has been a reluctance to develop technologies such as DAT audio into the domestic markets because it would be so easy for consumers to make perfect pirate copies. In spite of an agreement to assign a percentage of the income from selling tapes and equipment to royalties, there is at the time of writing no rush to release popular material in DAT format.

REVIEW

You should have learned the following things about the media as a source of communication and as Institutions, through this chapter.

1 DOMINANT CHARACTERISTICS

- the media are mainly characterized by the following aspects: monopoly, size, vertical integration, conglomeration, diversification, as multinationals, cooperation with one another, by control and domination of production/finance/product/distribution, particular values.

2 SOURCES OF FINANCE

- the media have access to enormous sums of money through their means of finance. These means are dominated by selling advertising space, by the sales of their main product and through the sales of associated products.

3 PATTERNS OF PRODUCTION

- there are distinctive patterns to the way in which media material is produced. These may be summarized in terms of routines, deadlines, slots, specialized production roles, use of new technology and heavy marketing of the product.

4 CONSEQUENCES

- there are major consequences of the way that media institutions are set up, run and financed. These may be summarized in terms of mass product, targeted product, repetition of product, elimination of unprofitable audiences, exclusion of competition, polarization of audiences and reduction of choice for the consumer.

5 MEDIA POWER

- the points already made lead us to conclude that the media have a great deal of power over what we read and see. The nature of this power may be discussed in terms of ownership, levers of power, power over product and ideology, the power of professionalism, cultural imperialism.

6 MEDIATION

- the media inevitably change the material which they draw on and shape it in certain ways.
- they construct this material and the meanings in it. They offer us representations, but not real life or real events.

7 CONSTRAINTS

- there are certain constraints on the ways that the media operate and so on the product which we receive. These may be summarized in terms of self-regulation through certain bodies, regulation of broadcasting through the Broadcasting Standards Council, of editorial control over the production teams, of legal constraints, and of financial constraints.

8 CENSORSHIP

- this is mainly internal and voluntary, and is described in terms of the constraints given above. External censorship is limited and operates through the law and through indirect government pressure.

9 MEDIA AND GOVERNMENT: RELATIONSHIPS

- there is a power relationship between media and Government. The Government exerts power over the media in various direct and indirect ways. The media need the Government as source of information, the Government needs the media as a means of communication.

10 PROPAGANDA

- this implies total control and deliberate use of media towards some end – which does not happen in this country.
- some talk of government influence on the media and of advertising, as something like propaganda.
- there has been propaganda in times of war.

11 FUNCTIONS OF THE MEDIA

- these may be summarized in terms of entertainment, information, culture, social and political functions.

12 MEDIA AND NEW TECHNOLOGY

- the arrival of new technology, symbolized by computing power, has changed institutions, production, distribution, product.
- the convergence factor, based on a common digital language, is bringing media and institutions together.
- access to this new technology raises problems of media access and of pirated products.

Activity (4)

This activity is concerned with the idea of institution – that one can partly understand media by understanding the organizations that finance, manufacture and sell the product.

This area is to an extent concerned with facts – about the technology of production or the mechanisms of distribution. But it is also concerned with the implications of these facts, not least in terms of the nature of the power wielded. So one may also look at censorship and at the empire of ideas which the media produce through their business.

Assume that you are involved with a possible deal to make a British horror film, set in a castle in Yorkshire, having two British stars and one minor American star.

ANSWER THESE QUESTIONS ABOUT ASPECTS OF PRODUCTION

- From where might you get the backing to make this film, and why?
- What kind of production company might actually make the film?
- Who would be a good distributor to go to to get the film screened and marketed?
- What kind of cinema/cinema chain would be likely to take the screening of the film?
- What are the main constraints that you can see, on getting the film made and shown?
- What would you *like* to do to make the project more attractive to a backer and to a distributor?
- How would you describe the audience for this film?
- What do you think you can *not* do in making this film, compared with a typical Hollywood movie?

If you can, work with two other people, or even form three groups, to cover three aspects of the film deal: production; finance and distribution; marketing.
The idea is that the production people make up a package which they take to the others. This package would include elements such as a title, stars, synopsis, key features of the production.
While the production people are sorting out the actual details, the other two groups need to make a list of questions to ask these people, which cover the main things which might worry someone in film finance and marketing.

PRESENT THE PACKAGE. ASK THE QUESTIONS. DECIDE WHETHER OR NOT THE FILM SHOULD BE MADE AND UNDER WHAT CONDITIONS

These activities should help you understand the way that media institutions operate. They are businesses with particular views of the world and particular concerns which may not be shared by their audiences, or by media students. The way that they operate helps explain what they produce and what meanings their productions may carry for the audience.

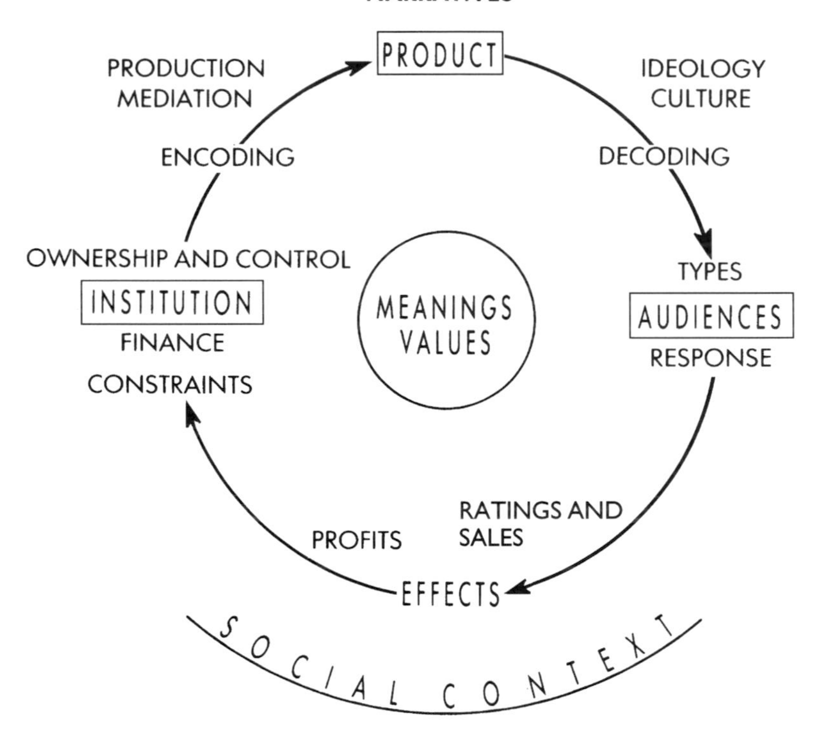
INFORMATION AND PERSUASION
GENRES REALISM
REPRESENTATION
NARRATIVES
PRODUCT
PRODUCTION
MEDIATION
IDEOLOGY
CULTURE
ENCODING
DECODING
OWNERSHIP AND CONTROL
INSTITUTION
FINANCE
CONSTRAINTS
MEANINGS
VALUES
TYPES
AUDIENCES
RESPONSE
RATINGS AND
SALES
PROFITS
EFFECTS
SOCIAL CONTEXT

5

Product and Treatment

Rolling Off the Production Line

Having dealt with the source of the message we are now dealing with the means by which it is carried. And just as the source was linked to the messages we must relate the media material to the messages, which are the subject of Chapter 7. The point of what is said, and the means through which it is said are inevitably joined together.

This chapter is about ideas which help make sense of the product. It looks at genres, stereotypes and versions of realism. It does not describe programmes or newspapers in detail, or how they are made. That may be interesting, and you may wish to do some follow-up reading (consult the book list). But what is most interesting is the SO WHAT question. We see many films or magazines coming out of the media institutions described and commented upon in the last chapter – but, SO WHAT? So – those many films, magazines and other **examples of product show some significant patterns when we look at them overall.** These patterns are significant because they suggest which meanings may be carried to the audience and how. Once more the meanings matter, the messages matter, if they shape our view of the world.

1 REPETITION

The first thing to notice is the sheer repetition of the production line material. **Much of the content and treatment of the output of the various media industries is basically the same.** This is most obvious in such examples as film sequels, television series, follow-ups and cover versions of records. Sometimes this is inevitable where there is competition for the same audience which likes the same things. For example, it is hardly surprising that every newspaper has sports coverage and repeats sporting stories. The popularity of sports can be seen in the number of people who go to these events. There is also evidence that television coverage relates to active involvement in sports, as with the expansion of snooker halls in the early 80s.

And the repetition – to take another example – of records across various radio channels, or even of various styles of music across these channels, must also signify something. Although it is slightly less obvious why there should be so many romantic stories both in women's magazines and in novels, the repetition of the formula is no less significant.

To put it at its simplest, **if something is repeated often enough it will tend to be believed.** If movies run to type, such as the cinema/video spate of films about Vietnam, then this gives them a high profile in public consciousness, and means that any messages they contain are likely to be correspondingly emphatic. This repetition is not limited to types of film or types of article. It includes types of treatment. Such **repetition of treatment causes it to be accepted as the 'right' way to handle the material**, the way in which one expects it to be handled.

For instance, the sheer repetition of electronic drum sound and of sampled records in one type of popular dance music at the moment has created acceptance of this treatment, and a willingness on the part of the audience to buy more of the same. The repetition of electronic special effects in films, especially science fiction, has created a similar acceptance and audience demand. Viewers expect to see convincing laser battles, just as viewers of TV news expect to see authentic footage of wars in other parts of the world. Indeed they expect to see what is happening – drama and action – with brief packaged explanations, rather than, say, photographs accompanied by longer background explanations.

There is nothing 'natural' about the way in which material is handled. We expect to see things done in certain ways as the result of years of repetition, which creates an understanding between the creators and the audience. If we are going to unpack the meanings in media material, then we must expose the idea of 'natural' for the con job that it is. We must get to the heart of the con (which we are all happy to join in!), by noticing the repetition and asking how and why it takes place.

The main reason why we have repetition of content and treatment is because it sells. All media operators want to make money, to attract audiences, to improve their ratings, to improve their readership figures – whatever is appropriate to the given media. The repetition may help acceptance of the content or form of the communication. But it still has to be based on something which is attractive to the audience. There is no way we can be persuaded to buy something that we really do not like or want. There is no way that we can be sold something that we really do not believe in. There is more about persuasion and effects in the final chapter. But for the time being you should accept that there is no evidence to support any kind of conspiracy theory about the media – that the gnomes of press and television invent items and ideas which they force upon us through some kind of fiendish plot. The truth is probably more insidious and of no less concern – that they pick up on things that are already there. **So the reasons for the kinds of repetition that we can identify have to do with material motives rather than intentional and ideological ones.** They are connected with the production routines that we have already considered. It is easier to do things the way they have been done than to change one's routines. Why change what you are doing or the way you do it if it is making you good money, so the argument might go. Of course content and treatment do change, people do think up new ideas and new approaches. Where there are new elements in media material, it is often because the one medium borrows from another, because someone is trying to improve on another person's idea. Computer games of 'Dungeons and Dragons' are borrowed across into a television children's quiz game *Nightmare* with computer effects.

So the causes of this repetition are mostly pragmatic, commercial, even instinctive. What is being repeated, remember, is any one of three elements:

- the categories of media product (e.g. type of programme)
- the content of that product (e.g. type of character)
- the treatment of that content (e.g. the device of reprising sections of film with freeze frames at the end of a TV film in order to give the actors' credits and more screen time).

As for **the consequences of this repetition**, it reinforces these three elements in the minds of the reader or viewer. More than this, it reinforces the covert messages embedded in the material. And this is why it matters. For instance, television has produced hundreds of hours of police thrillers. Many of these have male heroes who are mostly action men, with little sign of the personal anxieties and problems that dog most of us mortals. These examples give us the not very covert message about what it is to be a good cop and good guy. It is encouraging that some programmes which break this mould, e.g. *Prime Suspect*, have in fact been very successful. They show us that women can also be good cops and still deal with personal problems.

Mention of such a type of programme brings us neatly to a major concept and way of categorizing material in the media – genres. So let's look at what genres are and why they matter.

2 GENRES

2.1 Definition

A genre is a type or category of media product, like a spy thriller. It has certain distinctive main features. These features have come to be well understood and recognized through being repeated over a period of time. Sometimes there are variations on the genre, which may be (rather awkwardly) called a subgenre – the Bond spy thrillers or the space-opera type of science-fiction film. The term is generally associated with fiction material, though news could also count as a genre of television.

In some ways it is easiest to define genres by their generic titles – science fiction, cop thriller, Western, quiz shows. But this lands us in a chicken and egg situation. Why do we dredge up such a title in the first place? To say that genre is, for example, thrillers or that spy thrillers are a genre takes us round in a circle, and does not sort out what genre is.

The term does not necessarily cover all sorts of repetitive material. In particular, it **should not be confused with modes of realism or of style.**

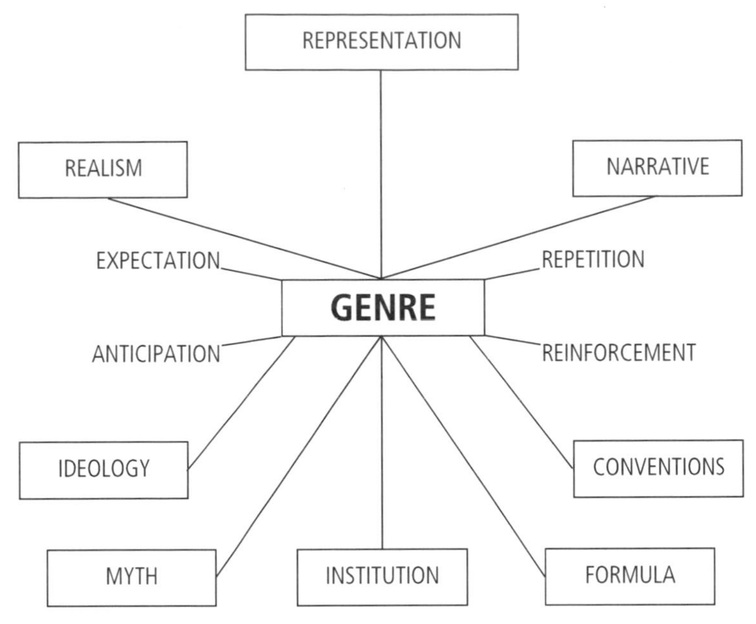

Fig. 5.1 Key concepts – Genre

Documentary or melodrama, for instance, are ways of handling the material, and as such cut across any genre. They may have certain repeated elements of their own – use of natural lighting or exaggerated confrontations. But they don't have any clear pattern of expected subject matter or of character. In other words, we could have a melodramatic cop thriller or indeed melodrama in any other genre. But not all cop thrillers are melodramatic. In the same way, other modes such as comedy and romance may have certain recognizable repeated traits, but they are not as predictable as genres. Any story, genre or not, may be treated as a comedy. But comedy in general is not sufficiently predictable to be seen as a genre.

These remarks suggest that there are further characteristics of genres, such as predictability and repeated elements, which make them distinctive and which help define them. We should now look at these.

2.2 Key Elements

All genres have a portfolio of key elements from which they are composed. Not all examples of a genre will have all the elements all the time. There may be permutations of these. It is these elements which make up the **formula** of a given genre. They add up to what we unconsciously expect to see and to enjoy seeing or reading. The way that the elements are put together is itself formulaic. As with other topics in this book, I hope to make conscious what you really knew all along, and to make some sense of it. You should also realize that any genre is more than the sum of its parts. So in what follows I will describe the parts separately, but you should understand that whatever the genre means to you comes as a result of your taking in all the elements together.

Protagonists

All genres have recognizable protagonists or lead characters. These may be heroes and/or villains – as in the film *Seven*. Sometimes these lead males and females are so predictable that they have the same qualities across a number of genres (see archetypes, below). Such males would be courageous and good looking, and likely to rescue a lady in distress at some point. By the same token, the lead female is likely to be very good looking, and play second fiddle to the male hero. A popular film such as *True Lies* exemplifies this tendency, though it is more extreme in action comics (*Doc Savage*) or cheap American TV thrillers (*Matt Houston*). It is also true that many of these male protagonists are loners, whether one is talking about, say, the Western or about the Private Eye movie. Sometimes actors invest such a character with some individuality – usually because they are seen to be fallible. But in many ways one would be hard put to truly distinguish one genre hero from another simply in terms of character and behaviour. The same could be said of quiz show hosts, who are also part of a formula and a genre. They are a kind of protagonist who isn't much different from one show to another.

Stock Characters

Another part of the scenario includes recognizable though minor characters, who sometimes are repeatedly played by actors who specialize in such a part. Walter Brennan used to be expert as the Old Timer in the Western, dispensing homespun wisdom and hitting a beetle with tobacco spit at 50 paces. The robot servant is a stock character in science fiction, all the way from Robbie the robot in the 1950s, through to C3PO in *Star Wars* in the 80s. Lovely Leila (or whoever) is a stock character in many quiz shows, as decoration and dispenser of prizes. The reporter on the spot is a stock character in news.

Plots and Stock Situations

The storylines or parts of them are also predictable and recognizable. Gangster films are almost invariably about the rise and fall of the gangster hero, and will include a final shoot-out scene in which morality triumphs and the anti-hero gangster dies in a hail of bullets. Soaps are just as predictable, however complicated the storylines become over weeks of viewing. There is bound to be a scene in which someone turns up from the past of the hero/heroine and has some kind of confrontation. Romantic stories, including the photostory example, often include a plot line in which the heroine cannot get the boy because of some awful obstruction to their love – perhaps simply the fact that he does not notice that she is there. But it will all come right in the end. Or it will turn out that he is a bad lot anyway and she is well off without him. Further examples of stock situations within the plots are plentiful, for instance the scene where a suspect is given a grilling in police thrillers.

James
LEE BURKE
THE NEON RAIN
"His is a name to watch"
Los Angeles Times
(a)

AGATHA
CHRISTIE
The Mirror
Crack'd from Side to Side
HAS THERE EVER BEEN A MURDER WITH A MORE INTRIGUING MOTIVE?
(b)

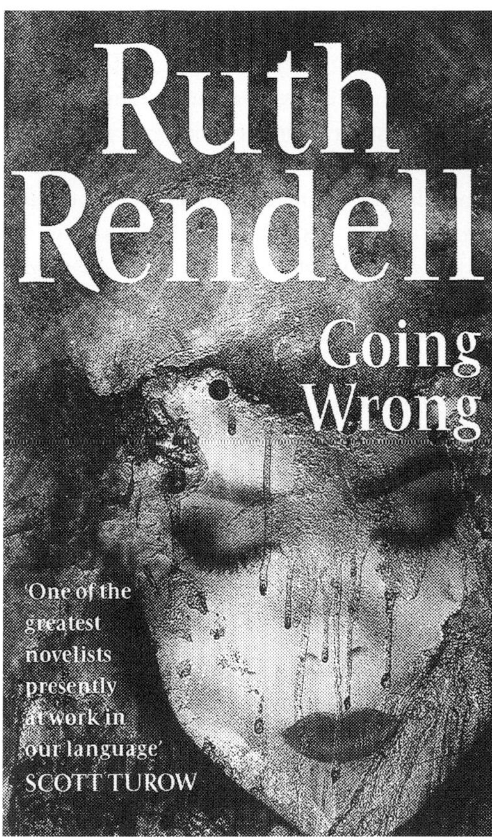

(c)

(d)

Fig. 5.2 Iconography of the thriller

The icons on these book covers provide an instant point of contact for the target audience. Of course there is more to the cover designs and their meanings than just the image of a dagger, for example. Equally there are similarities between the covers which place them firmly within the genre. These are all points which you can tease out.

Icons

This element is crucial to genre because, above all, it is the aspect of the genre which we immediately recognize and lock into. There are three main types of icon – objects, backgrounds and sometimes stars. The icon is a key symbol of the genre – see the icon and you know immediately what territory you are in. All the other elements of the genre are likely to be assumed once you have recognized the icon and interpreted it. Icons such as the Colt 45 of the Western, the Luger of the spy thriller, the ray gun of science fiction, are potent symbols. They do not just stand for the other genre elements, they also stand for the main ideas and themes of the genre. A starry background with planets stands for adventure and exploration. The titles sequence of Independent British television news has used the icon of Big Ben for years. It stands for integrity and authority – ideas which the programme continues to project in the way it presents its material. The faces of the lead actors and actresses in TV soaps may also be said to become iconic. This emphasizes the point that communication is learned. It takes time and repeated exposure to examples of genres before new elements or new icons, are created in the minds of the audience.

Icons are generally visual elements. However, they can appear in other forms. For instance, the whine of a ricocheting bullet is fairly synonymous with the Western. The squeal and chatter of computer terminals is fairly synonymous with science fiction. It is much harder to apply this concept to print media, because there is not that quality of instant recognition – elements take time to unfold. However, icons such as the distinctive uniform of the American cop may be described in words and still be recognized as special to the genre.

Backgrounds and Decor

These elements are also typical, distinctive and recognizable for a given genre. Their importance varies from example to example. The interiors of a current soap such as *Neighbours* are pretty anodyne. But the pub interior of *EastEnders* is distinctive and a part of the character of the series. A shot of rain-slicked city streets at night with high-rise buildings is likely to lock one into either the private eye or the cop thriller. If such backgrounds become very distinctive then they become icons. Some interiors are most recognizable in terms of their decor (the furnishings and their style). The glossy tinsel and flashing lights decor of many quiz shows is distinctively part of their genre. So are the control room, consoles and screens of science-fiction interiors.

Themes

The themes or ideas which run through and come out of the stories are very much part of genre. Some are relatively common to genres, some are more special. These themes also tie in with the value messages which are projected. For example, all genre narrative says something about conflict between good and evil, between alternative views of right and wrong. But the theme of deceit and betrayal is more special to private eye stories, where the hero is often double-crossed by a female. The theme of materialism in American soaps (acquiring money and possessions) is at the same time a message about the desirability of having these things. By making something a central idea in a story one also makes it important. The theme becomes a message.

2.3 The Formula

So all these elements add up to something called the formula. They are if you like headings for the kinds of item one would expect to see in all genres. When you look at particular genres then of course as with the examples above, it is particular items which you describe for that genre.

For example in *Television Soaps* (1992) Richard Kilborn starts off talking about the characteristics and the formula of soaps in terms of the 'never-ending story'. He also points to the complexity of variations on the formula of any genre when he says that although the audience has a considerable 'fund of knowledge as to what soaps are and how they work' their main concern is usually with 'the individual model rather than with abstract questions of design.' However, media students ought to be concerned with the design and what we can learn from it.

So in the case of soaps we can list other likely elements of their formula such as matriarchal figures, family-centred drama, conflict through misunderstanding, emotional crises, locations which act as a focus for the community, a range of generations in the characters, episodic structure with cliffhanger endings, parallel multi-stranded narratives.

The audience's knowledge of the formula helps know what is going on and gives the pleasure of feeling on familiar ground. And if it gives the audience a blueprint for making sense of the drama, it also gives the production company a blueprint for making the drama, one which it can feel confident it shares with that audience. It is good to recognize the existence of a formula. But genre study is not just about enumerating elements. It goes on to be about why that formula matters in terms of the meanings it inscribes into the text, and the meanings which the audience decodes from the text.

2.4 Recognition and Attraction

So genres have this quality of being very recognizable for what they are, sometimes from just one or two shots or pictures. Indeed **the story makers depend on this recognition for instant communication with the audience**. The great thing about genre material, from their point of view, is that no time is spent setting up a character or explaining a situation. If it is familiar then the audience know the kind of person or scene that they are dealing with. In fact the story makers can trade in recognition in order to tease the audience, by doing something which they do not expect.

So the attraction of genre material is, among other things, the **mixture of familiarity and the unexpected**. The audience will run the video of *Nightmare on Elm Street 5* partly because they already have had the pleasure of being

	Total interested[1] %	Very interested %	Quite interested %	Not that interested %	Not at all interested %
National and international news	85	48	36	10	4
Local and regional news	85	41	44	10	4
Films – recent releases	81	42	40	11	6
Nature and wildlife programmes	76	44	32	12	10
Adventure or police series	70	25	44	17	12
Plays and drama series	68	21	47	16	14
Situation comedy shows	66	18	48	19	13
Soap operas	61	32	30	18	19
Crime reconstructions	59	21	38	19	19
Sports programmes	56	33	23	19	24
Holiday and travel programmes	56	18	38	22	20
Health and medical programmes	55	16	39	23	20
Quiz and panel game shows	54	22	32	22	22
Older or classic cinema films	53	22	32	20	25
Hobbies and leisure programmes	53	12	40	28	17
Current affairs programmes	52	13	39	21	24
Chat shows	52	12	40	24	22
Variety shows	50	16	34	25	22
Alternative comedy shows	48	15	33	22	27
Films suitable only for adults	42	12	30	24	32
Programmes from or about European countries	42	7	35	25	30
Consumer affairs	42	7	35	26	30
Science programmes	39	12	27	23	35
Women's programmes	38	11	26	21	38
Education programmes for adults	37	11	27	28	31
Pop or rock music	36	14	22	18	43
Arts programmes	23	7	16	31	43
Programmes about politics	22	5	17	26	49
Programmes for older children	22	5	17	19	55
Church services	21	6	15	19	57
Business and financial programmes	19	4	15	26	52

	Total interested[1] %	Very interested %	Quite interested %	Not that interested %	Not at all interested %
Programmes about religion	18	3	15	22	57
Programmes for the under 5s	17	5	11	14	66

Base: All TV viewers. Notes: 1. Sum of 'very' and 'quite' interested (may not sum exactly due to rounding). 'Don't knows' excluded

Fig 5.3 TV genres and TV viewing – interests in types of programmes

Figure 5.3 gives a different angle on the idea of genre and popularity. It shows that people say that they are most interested in news, even though it is soaps which consistently make the top rating of what people actually watch. Similarly one could reflect on the fact that although sport has a dominant position in programming (not to mention the content of newspapers), still one fifth of interviewees said that they weren't interested in it.

frightened out of their socks, and partly because they expect something a bit different, without knowing what, exactly.

2.5 Anticipation, Expectation and Prediction

All media material gives some kind of pleasure to the reader or viewer. This is why the audience buys the product. The nature of that pleasure is another matter. Obviously, the kind of enjoyment we get from reading a newspaper is not the same as that from reading a novel. Within genres specifically, there is a range of pleasure to be gained, not just some sort of gut excitement at seeing Schwarzenegger take on the Predator or whatever provides the excuse for violent special effects in that kind of movie. There is also a kind of pleasure gained from being able to anticipate what will happen next. This is most obvious in the mystery thriller genre, where one of the main points is to second-guess the identity of the criminal – *The X Files*, for example. There is pleasure to be gained from predicting what will happen. People discuss soaps such *Brookside* in terms of the motivation and likely behaviour of the characters – 'I reckon she's going to run off with him' – and so on. There is the pleasure of expectation realized. The reader's trembling hand turns the page of the comic, expecting to see the hero get out of a tight corner, probably through another display of martial arts, whether it is *Streetfighter* or *Slane*.

2.6 Seriality

This notion refers to the way in which media material, especially genres, repeat and rework story elements. Berger (1992) discusses and explains the ideas of Umberto Eco on this subject. He describes a four-part typology as follows.

- **The Retake** – where a new story is built around characters from a previous successful narrative. An example would be *Star Wars*.
- **The Remake** – where a known story is retold with variations in an 'up-to-date' version. An example would be *Dracula* or *Wyatt Earp*.
- **The Series** – where a new story or set of stories is made around one central character. Detective series in all media are a good example of this – say, *Inspector Morse* on British television.
- **The Saga** – where the story is built around the development of a family over a period of time. Many soaps are examples of this. Examples would be *The Archers* on radio, or *Neighbours* or *Dallas*.

This is a useful description which draws attention to key features of genres, such as conventions, repetition and the formula.

2.7 Repetition and Reinforcement

This topic is worth returning to. The whole of genre depends upon it. **The building blocks of genre, its elements, as well as the messages that genres communicate, all depend on being repeated, so that they continue to be known and understood by the audience.**

To some extent, a genre can become self-perpetuating once its key elements are established. The more the stories use the same elements, the more the audience accepts that this is what the genre is all about. But genre is not static, it is always adding variations to its elements and its formula. The whole invention of *Star Trek*, and its transfer from TV to film is a good example. But **the meanings, or messages, are also repeated and reinforced. They become 'natural'**. They are believed the more they are 'said'. The dominance of white ethnic groups across the entire range of genre material in the multiracial US and UK clearly carries a message about 'white rules, OK'.

2.8 Formula, Code and Conventions

When we recognize and make sense of genre material we take all the elements together. This combination of elements special to a genre represents a kind of formula. This formula may vary slightly from genre piece to genre piece. But essentially it is there all the time, composed from the elements that have been described.

Each **genre also represents a kind of special code, shared by its makers and its audience**. In terms of semiotics, or the study of signs and their meanings, we are dealing with a secondary code (see Chapter 3). For example, a film musical is based on all the primary codes of speech, non-verbal communication, music, and so on. We have to know these in the first place in order to make any sense of it at all. But there are special qualities to the musical which overlay a secondary code on the primary ones. For instance, it is a part of the code that people may break into song at improbable times.

So **there are various sets of rules for what should be in a code and how these elements should be used and combined**. For example, there is a general rule for chat shows on television (as with light entertainment programmes) that hosts and guests should make their entrance down a staircase. **Such rules are called conventions.** Most programmes, whether they are genre or not, reveal some conventions in the way that they open. You should try an evening's viewing devoted to the beginnings of programmes and see how many rules of the game you can spot.

The idea of convention is double-edged in the sense that it can draw attention to what is similar when comparing texts, or it can highlight what is different. This is just as important because what keeps a genre fresh and popular is the ability of its producers to invent variation and difference, without losing the security (for the audience) of a degree of sameness or repetition.

There is also an interesting point about the association of genres with popular and therefore supposedly lowbrow culture. There is an opposed assumption (a dichotomy) that original, unpredictable non-genre material is somehow more highbrow, culturally superior, simply better. This valuing of that which is special and unique is rather peculiar to a Western tradition. It is allied to the example of the status of the novel, created by the lone artist, and not in a genre. There is no absolute logic to this position. In India for example, endless retellings of the stories of the gods, whether in film or in puppet theatre, are not devalued in the minds of people simply because they are a genre.

2.9 Genre, Industry and Audience

Figure 5.4 makes it clear that there is indeed a tight relationship between the media industries that manufacture genre products and the audience which

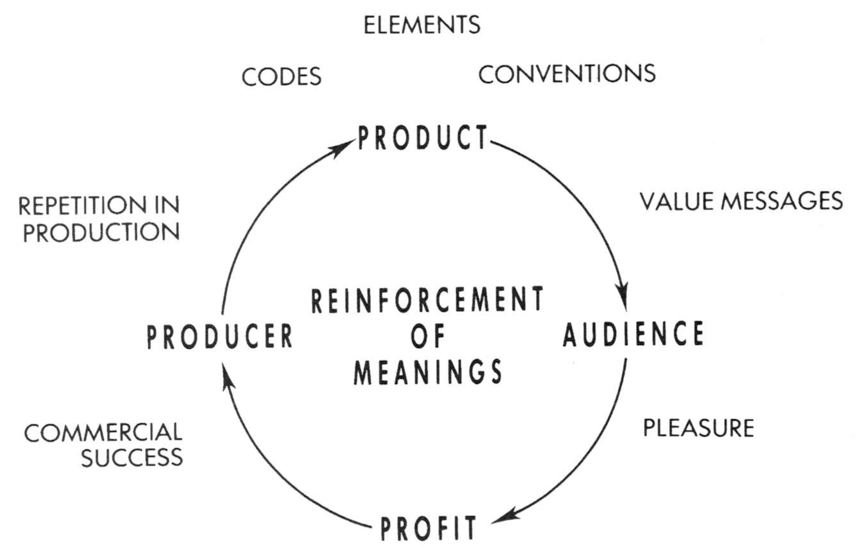

Fig. 5.4 Genre: a model describing key elements

consumes them. Genres are good for industries because they are generally good for profits. They are good for profits because, by definition, the audience pays for them consistently, and sometimes very well. The audience is attracted to genre material and pays for it because it takes pleasure and satisfaction from the material. The producers are also happy to make genre material because, being familiar, everyone knows how to handle it, and because they may literally be able to reuse sets and props. In television, the soap genre is of special economic importance because a successful soap provides a secure base on which to build an evening's programming, and a secure base on which to build a company's fortunes (*Brookside*).

The crucial entry point in the circle of profit and pleasure is the nature of this pleasure. I would suggest that there is something here which does not seem to be about pleasure at all, on the surface. This ties in with the messages which the audience obtains from genre material (indeed all media product), whether consciously or unconsciously, and perhaps the important dimension is to do with confirmation of beliefs. Genres tend **to confirm what we believe and what we want to believe**.

This links with ideology once more. It ties in with a suggestion of Feuer (1986) that we can categorize genres in three ways. One is **ideological** and refers to the idea that genres are distinctive in the way that they incorporate values of the dominant ideology; indeed that perhaps some genres incorporate values more or less specific to themselves. For example, most police detective stories endorse the position of the police person and the law. A second categorization is described as one of **ritual**, which is about the way in which genres distinctively cycle around between industry and audience. The present 'fashion' for serial killer movies is a particular example of this. The last category she calls the **aesthetic** and is about textual characteristics – aspects of form. This I have described above as key elements of genre.

2.10 Genre and Myth, Genre and Culture

These beliefs are part of our culture. They may be special to our culture. For instance, soaps deal with the idea of true love. Sometimes they send the belief message that true love is more important than material goods, and perhaps more important than marriage (see the theme of infidelity). Many other cultures would reject both of these ideas out of hand. They would not even put them up for discussion in their own genres.

And you should realize that other cultures do have their own genres, which have exactly the same main elements as have been described. The Japanese enjoy Samurai stories – historical adventure dramas. We would find it a little difficult to make sense of them because we don't know the conventions. We would certainly not appreciate the full importance of their messages about respect, ritual and loyalty.

All media material is inevitably a product of the times and the culture which makes it. It is arguable that **genres have a special place** in this respect. This may be for two reasons. One is that they carry their messages in the protective wrapping of an established popular form of entertainment. The other is that they are based on core topics which if not universal, at least don't age quickly. The crimes of thrillers, the family turmoil of soaps, the science of science fiction is likely to be with us for the foreseeable future. The war film, for example, may be 'used' to say things about real wars which concern people at a given period of time: *Tumbledown* on television, for the British and the Falklands experience; *Gardens of Stone* in film, for the Americans and the Vietnam experience.

But of course genre is not only based on real events, or on factual events of history. **It may be based on versions of that history, or even on nothing more than myth and legend.** (Myths are stories which have no basis in fact, but which nevertheless represent some kind of truth for their culture.) The story of Billy the Kid has been rewritten many times through the Western in comics, novels and films. In general, those versions have become myth, because people wanted to believe that he was a genius with a gun, that he helped the poor, and so on. The truth is very dull and rather unpleasant. But many Americans want to believe in the individualist as hero, in the rightness of winning the West. In the same way, the Dracula horror story has been remade many times in all media of communication. The Bram Stoker novel is the genesis of this story. But this is almost irrelevant to the fact that audiences want to deal with their fears of death. They are fascinated by the idea of being immortal, yet feel one should be punished for becoming immortal like God. Hence the creation of the Dracula figure. As a piece of myth, it works through themes and anxieties within our culture.

2.11 Intertextuality

The notion of **intertextuality** is not peculiar to genres, but it is strongly exemplified by them. What it refers to is the way in which we understand one text by reference to others. The links between texts operate in many ways, the most basic of which will be the visual and verbal languages which they share. On one broad level you could say that all comics are understood by reference to one another, even to all other texts. We understand the world with reference to everything that we have experienced. But intertextuality really makes sense when one gets down to specifics. For example one understands a romantic kissing scene with reference to all other such scenes.

Other links are those to do with borrowed techniques or with allusions in one text to another. For example a current series of advertisements on TV for Lloyds Bank makes light-hearted allusion to fairy stories, and also uses the technique of having a voice-over storyteller.

Genres are understood intertextually through others of the same genre, but also through other genres. So, in a particular sense, one understands aliens in science fiction by cross-referring one story with another, through our knowledge of other stories. In a general sense, within genres, one understands aliens with reference to enemies in war films or creatures in horror films, for example. And, in the even more general sense of intertextuality, one understands aliens through reference to stories about travels in foreign lands and about strange creatures, whether factual or fictional.

Another way of recognizing intertextuality is in parodies. We understand the jokes through our knowledge of other genre texts. Parody reinforces the very existence of a genre because if we did not 'know' about the genre and its features it would not be possible for the author to make fun of them, knowing that the audience understood what was being referred to. As Lorimer says (1994) 'intertextual connections are part of the taken-for-granted knowledge, understandings and competences used by consumers of popular culture.'

2.12 Case Examples and Value Messages

I would like to round off this section by having a quick look at some specific genres in various media. In particular, it is worth having a look at the dominant characteristics and value messages present within each genre. Remember that all fiction material now travels freely between film and TV media. All films are shot with the TV aspect ratio (screen proportions) in mind.

Film and TV Cop Thrillers

Television helps keep this genre alive through its appetite for more and more material, so that new variations appear every year – *The Bill* on British television or *NYPD Blue*. Films like *Serpico* are made for cinema and TV series follow.

The ability to create action sequences and to go into real or simulated locations gives a particular edge to this genre, which is also fed by documentaries about police work and through crime-solving programmes such as *Crimewatch*. It has a sense of being contemporary and immediate. Crime is a topic in daily newspapers. The genre concentrates not merely on the cop hero as crime solver, but on the nature of the crime and on retribution to the criminal. The cop thriller genre legitimizes the depiction of criminal acts, which has its own fascination. It also must deal in fundamental value messages about right and wrong.

Indeed, the particular interest of such genre material is that it raises questions about when the Law is effective and when it is an ass; about when police are fair and when they lapse into criminality themselves; about who should be punished for what, and how. It raises questions about the nature of punishment itself. Such questions, or issues, are central to ideology because ideology is about the exertion of power and about the beliefs and values which are behind that exertion of power. Beneath the action and the human drama, cop thrillers are very political stories.

Magazine Romantic Stories

It is important to distinguish between these stories and romance as a mode of treatment of stories in general. These stories do contain repetitive elements as described above, whereas romance in general may appear in almost any other kind of story in any other medium in any other genre.

Such romantic fiction centres on a heroine, with strong supporting roles of female rival and best friend, and the male romantic object. It has well-defined plots and stock situations involving romantic encounters, romantic conflicts, and obstructions to romance. Its iconography is least well defined. So far as it is distinctive, it relies a great deal on the exchange of looks, the embrace and the kiss. The backgrounds are often historical (poor girl in love with rich man) or exotic (she met him on holiday in a foreign place). There is also a strong vein of background realism (girl next door falls in love with home-town boy). Again, backgrounds are not strong in iconographic terms – once more weakening this as a distinctive genre.

The dominant themes or messages are about the value of romantic love, often tied to the value of marriage as a means of sealing that love (and ending the story!). It is significant that romantic stories for young females rarely engage with life after marriage or with serious issues of personal relationships. All that matters is love itself. Associated messages are about the importance of being the centre of the loved one's attention, about being misunderstood by parents and/or friends, about being recognized as an attractive and loving person.

■ *Television Quiz Shows*

Because genres are dominantly fiction, this may seem at first an unusual example. But in fact it works, if one looks at quiz shows in detail. Indeed it shows the power of packaging media material, and the convenience of typing it for producers and audience.

Quiz shows are a kind of story. If we concentrate on the subgenre of game shows, then the compère is a kind of narrator who takes us through a familiar story of competition and success. The participants are the heroes and heroines. The obligatory female is one stock character; the studio audience is another. The icons are the well-known compère and the sets with flashing lights, display boards and the like. An example of a stock situation is when the competitor has to make a decision as to whether or not they will go on to the next stage of the game, take the money on offer, or whatever.

These stories also have value messages attached to them. Remember that these are the whole point of examining and describing the characteristics of this genre. These messages are about it being OK to compete and to aspire to material goods. They are about turning the consumption of goods into fun. And there are other messages which have to do with stereotyping (see Section 3). The treatment of the female 'hostesses' as objects is a fairly obvious example. Less obvious examples come through the treatment of the participants by the compère. Usually they are talked to in such a way that they are firmly subjugated within the television system. It is as if that system does not want the participants to get above themselves: they must stay within the rules laid down by the programme. I have even heard remarks made about the national dress of ethnic minority participants. These remarks clearly conveyed messages about what was to be seen as 'normal' and 'dominant' in our culture. They diminished the participants. They were essentially racist.

Television News

This is another overtly non-fictional genre which looks more fictional the more closely one inspects it.

Again, we have an omniscient narrator, who might also double as hero, steering us through the events of the day, coping with news coming in through the earpiece, bringing the drama to a successful close with the final summary and amusing tailpiece. Stock characters in this drama are the reporters, star subjects of world events, the experts and the eyewitnesses. The main plot of the news has its highs and lows of drama like any story. The individual stories are themselves dramatized to a greater or lesser degree, having qualities of conflict and suspended endings so that we watch the next episode to see who will win. Stock situations exist because the news has its conventions of content and treatment like any genre. These conventions add up to the same thing as news values. For example, what is valued as a rule is disaster, major political decisions, deaths of the famous, royal activities – you can add to the list for yourself.

The **dominant messages** of this genre are firstly to do with the authority and integrity of the news organization itself. This is essential if it is to be believed and trusted. Hence conventions of treatment such as face to camera, actuality film footage, conventional-looking newsreaders and reporters. But secondly, the messages are to do with the kind of world that we live in. These messages very much support what is called the **dominant ideology**, that is, the view of the world held by those people in our culture who actually run things. The truth or importance of these messages is another matter. What gives cause for concern is the way they are obscured by being naturalized. As viewers we should be able to decode the communication completely so that we can decide whether or not to accept such messages. On one level we should be able to stop and decide whether or not we really want valuable news screen time occupied with the activities of a wealthy middle-aged lady called the Queen. Maybe she isn't that important. On another level, we might ask ourselves whether messages about the importance of increasing industrial output, of having continued economic growth, are also valid. They may fit one view of the world, of a certain kind of capitalism. But there is nothing intrinsically good or right about perpetually producing more goods. It might be good to have a situation in which no one gets more of anything, but in which our culture improves the quality of what it has got. These comments are not meant to represent any one view as 'right'. Nor are they meant to suggest that all news is biased or completely lacking in alternative views on issues and events. But they are meant to question the notion of being right, which is at the heart of news messages about our way of life.

3 REPRESENTATIONS (STEREOTYPES)

Just as the material which rolls off the production line is dominated by genres, so too **there are dominant representations of people in the product**. In fact,

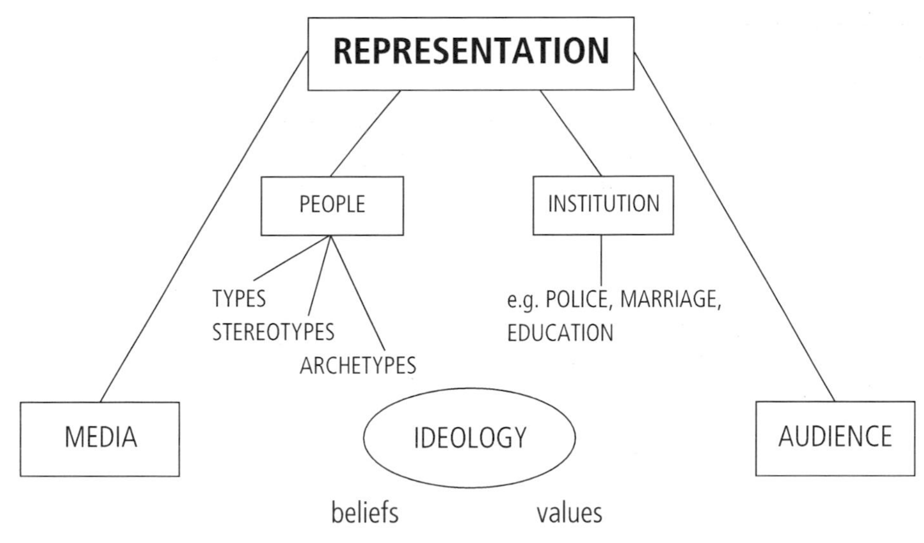

Fig. 5.5 Key concepts – Representation

many of these representations appear in genres, though not exclusively. For example, newspapers represent groups of people in certain ways. Advertising – which is a way of using various media – certainly does this. And even pop music may give us such representations through its videos and possibly its lyrics.

So once more, look at the media as a whole, if you can, because the representations of women in general or students or working people are put together across the media and understood by us through all the media. If you think this sounds like more repetition and reinforcement, then you are right.

What is represented is certain views of these social groups. And it is these views which we unconsciously learn to accept as normal – to the exclusion of alternative views. Too often such views are negative. In his book *Hiding in the Light* (1988), Dick Hebdige points to the ways in which young people are represented in terms of 'confrontations, of consumption and of life style'. So often they exist in media texts only when they are a problem. News media and even books like *A Clockwork Orange* are 'guilty' in this respect. These meanings start with representation of appearance and of character.

3.1 Representation by Type

You may think that this section is really just about stereotypes. But people can be represented in certain ways through certain devices without becoming actual stereotypes. So we need to define our terms carefully. I would suggest that when it comes to putting people into categories, there are actually three levels at which this happens. At each level the representation becomes simpler, cruder, more generalized, more clichéd, more worrying in terms of the value messages underlying what we see or read.

Types

At the most general level we can talk about something called a type. We recognize a category of character in a story, such as the shopkeeper type. But for various reasons this character does not emerge as a stereotype. One reason may simply be that they are not drawn in very strongly. It may also be that they lack a clear set of characteristics reinforced by years of repetition, which mark immediate stereotypes such as the fanatical German officer in a war comic. Again it may be that, while the character is a recognizable type in a story (the eccentric old lady who solves crimes over her knitting), the fact is that they are actually drawn in some depth.

Stereotypes

The true **stereotype** is a simplified representation of human appearance, character and beliefs. It has become established through years of representation in the media, as well as through assumptions in everyday conversation. It is a distortion of the original type because it exaggerates as well as simplifies. It has qualities of being instantly recognizable, usually through key details of appearance. It has attached to it implicit judgements about that character (covert value messages). Stereotypes are much like icons of genre in that they are recognizable and they do carry along ideas. Stereotypes are not necessarily bad in themselves – it depends on how they are used and what value judgements they unlock. For example, a 'safety in the home' advertisement including a comfortable granny in a rocking-chair, which is used to make children aware of the dangers from electric sockets in the home, could be said to be OK. It represents granny as kind and careful. It is being used for a socially approvable purpose. Whether all older females wish to be seen in this way is another matter, of course.

Archetypes

The most intense examples of types are also very deeply embedded in our culture. They are the arch heroes, heroines and villains who epitomize the deepest beliefs, values and perhaps prejudices of a culture. Superman is an archetype, just as all those heroes from mythology are archetypes. Archetypal characters in genres really belong to all of them, and are not entirely special to any one. For instance the power-hungry villain who wants to destroy the world might turn up in spy thrillers, science fiction or horror stories. Qualities of courage or beauty, goodness or evil, are drawn most firmly and simply in these archetypes. They are the stuff of comics (*Judge Dredd*), of cheap television (*Buck Rogers*) of low-budget film (the *Mad Max* series). They may be enjoyable in the story, whatever the medium, but they also take us into realms of fantasy.

3.2 Representation and Construction

We have said that groups of people are represented in certain ways through the media. Indeed, we have said that this representation also helps create the idea that certain people belong to certain groups – a chicken-and-egg situation. So it seems that **the media organize our understanding of categories of people and about why certain people should belong to certain categories.** These categories become part of our thinking process – which we use to judge people in the real world as well as in the media. Such categories are called perceptual sets. So representations of people in the media help build and maintain these perceptual sets.

The representation of types has to be constructed from something. Just as a genre is made from the building blocks of key elements, so too **a type is composed from certain elements.** In exactly the same way as genre, the elements become familiar the more they are used – repetition and reinforcement. In the first place, these elements are those of **physical appearance** – hair, clothes, distinguishing features. In fact some people think that this is all that representation is about, and so completely miss the point about meaning and value messages. **Typing takes place by age, by race, by occupation, by gender.** Let us consider the elderly Oriental male. This wise cliché will of course have a wispy beard and moustache, long grey braided hair, and something like a gown to wear, as well as an inscrutable gaze. He has a long pedigree, all the way from the *Rupert Bear* stories to the rather more cleanshaven version in *The Karate Kid* films.

But this type like others is also constructed from **certain behaviours, actions and relationships.** These other elements of construction may appear in our example as dignified restrained behaviour, acting according to a sense of honour, perhaps using magic or so called Oriental wiles, and lacking relationships other than of master to pupil.

Representations are also constructed through the medium used. That is to say, there is a written or visual language which tells the story and so puts together a certain type and a certain attitude towards that type. If we consider television or film, then it is, for example, obvious that close-ups on physical attributes are used to draw attention to them, and so they cue us into the type which is being built up. To talk about this aspect of construction is to describe how the type is constructed rather then what it is constructed of. In the case of our Oriental, a typical close-up would be a reaction shot, drawing attention to the character's impassive expression and reflectiveness when faced with a problem.

3.3 Representation and Meaning

What is represented through these types is far more than a view of categories of people or of what they commonly look like. We are also seeing **a representation of attitudes towards that type.** Because they are constructed with certain characteristics and treated in a certain way in the story, **we are told implicitly what we should think of them.** We are being told what the type should mean to us. We **are being given a set of value judgements** because we are decoding value messages behind the surface representation. So in the case of our Oriental, we are told that he is wise yet alien, mysterious but dangerous. It is a small jump from the wise old Chinaman in *Gremlins* to the

TYPING/STEREOTYPING (image of old person)

Construction: physical elements

- The chair and table are of an old-fashioned design.
- The male figure is wearing slippers and glasses.
- The male figure has white hair and is balding.

Construction: actions/implied relationships

- The male figure is seated in a passive position.
- There is a small girl seated on this person's knee: likely to be a grand-daughter from her gaze and expression.

Anchorage

- The words 'retiring' and 'pension'.

Deconstruction: apparent meanings

- This is a picture of a retired grandfather with his grand-daughter on his knee.
- Old people look like this and want to do this sort of thing.
- Grandchildren are born shortly before retirement at 65 (males).
- People of 65 or more will be seen sitting around.
- People at age 65 show distinct signs of physical ageing.

Deconstruction: value messages

- Retirement as an old age pensioner should be a passive experience.
- Retirement should be given over to one's family.

Comment

You may or may not agree in principle with the meanings and messages. Nevertheless they are there. If you agree that the elements of construction (and the meanings) are repeated in various media then you will agree that this image is of a type or stereotype. You might agree that not all old people would want to be seen in this way.

Fig. 5.6 A stereotype: with analysis of key features

Fig. 5.7 From *Police, Camera, Action* and *The Bill* – representations of the police

The two images of the police from one fiction and one factual programme both contribute to the total of all such representations (not to mention to representations of law and order). In one still, a real policeman is attending an incident on the motorway, involving a burning lorry. In the other, a fake policeman is questioning a boy in the street. You couldn't tell for certain that one image is from a fiction and one from a documentary – something to discuss in the first place. But then one also has to consider how the police are represented overall, in terms of what they do, what they are like – in the end, how we are to regard them. These policemen are out on the street, involved in drama, dealing with people, on active duty. The police I know actually have quite a lot of dull repetitive work to do. But this is hardly represented in the media. You can take it from there in terms of adding to what you believe is said through the media about police work and the law.

scheming Emperor Ming in *Flash Gordon* – an archetypal villain if ever there was one!

All types carry meanings of one sort or another, which may be critical of or demeaning towards the category of person so created and represented. One favourite type in comedy has been the mother-in-law – unattractive, strident, demanding, in battle with the relevant son- or daughter-in-law. Individual examples representing this type may vary in treatment. Only some may be thought stark enough to be considered downright stereotypical. But if we take all the examples together then what is represented is more than the appearance and behaviour of this female, it is also what this behaviour means. It may mean, for instance, something about male fears of strong-minded females, especially those who cannot be dealt with slickly in terms of their overt sexuality.

In effect, there are elements and devices which construct the type. These make up the surface representation. But then these elements taken together suggest meanings about the type, which also become part of the representation (see Figure 5.6). What is most important is to realize that **types are about values as much as physical attributes**. It is one thing to describe a type of person, and to note that such descriptions come repeatedly through the media. It is another to get into interpretation, to deal with the SO WHAT question. So we learn to value or devalue certain types of people in certain ways. We learn this, it may be suggested, because of the repetitious drip feed of media material overall (see the final chapter, on effects).

3.4 Absence Means Something Too

When media material is put together things may be left out by design. With regard to people and types, it is worth realizing that meanings about groups of people are created by omission. In one study of American magazine fiction it was pointed out that, while ethnic minorities formed 40 per cent of the total population, they only appeared in fiction as 10 per cent of the characters. In this country, ethnic minorities form just over 3 per cent of the total population. But even this small percentage has been conspicuous by its absence. Whilst we can point to the racial mix of a soap such as *EastEnders*, we can also point to the fact that, for instance, Asian minorities are absent from sitcoms. If one misses out groups within the population then one is saying things about their lack of importance, and about the relative importance of those who are put in.

4 REPRESENTATION, CULTURE AND MEANINGS

When the media represent groups of people, they are often saying things about culture as well, because those groups of people may belong to a particular culture or subculture. For example, the media in effect tell us a great deal about US culture and the American way of life. They also tell us about our own cultures and subcultures – youth, the Scots, Northerners. Whether what they tell us is the truth is another matter. But some view of these cultures is represented. There are messages about them.

One problem in deciding what these messages and meanings are comes from the fact that there is such a range of media material. A lot depends on what the audience chooses to view and read. On the one hand, if you took in documentary and current affairs material in broadcasting, as well as the quality press, then you could have a fairly wide-ranging view of the USA. On the other hand, if you only view American TV thriller series, read the tabloids and popular magazines, then you will have another view, which could be described as limited. This view puts white American males up front, emphasizes violence, shows far more of the city than of the countryside, is mainly concerned with issues of law and order, and so on.

So not only are the media selective in the views of culture which they show, we the audience are selective too in what we choose to take from the media.

Having recognized this, it is still fair to comment on the ways in which the media do indeed define our views of culture and of cultural groups in particular. For instance, Britain, like most countries, tends to have one consensus view of its 'main' culture, with views of separate regions measured in some contrast to this. That is to say, all its national media are based in, and reflect a consensus view out of, London and the South.

This may be comparable to a centralized Parisian view of France contrasted with views of the Bretons or of the southern French. In the USA New York and Los Angeles may define the centre, as compared with the South or the Midwest.

British Northerners (or Bretons or Midwesterners) are rightly recognized as a subcultural division. What may not be right is the way that they are represented. Our Northerner, as perceived through television comedy or magazine articles or radio drama, frequently comes across as plain speaking, living in a relatively poor urban environment and much occupied with sports such as football and rugby. You can eleborate on this view of the Northern type. It is symbolized by fictional media characters such as Andy Capp in cartoons, or the middle-aged heroes of the TV series, *Last of the Summer Wine*. You can argue about the details. But the fact is that messages are projected about the Northern culture as much as the Northern type. This culture is defined by such media product. It is indeed defined for Northerners themselves. We see and hear little or nothing about young people in the North or about fine arts in the North nor even about economic success in the North.

We can extend this kind of analysis of cultural definition to all kinds of groups and areas. We may ask ourselves what view is given us of Muslim culture in Britain or of rural life. Then we may ask where our views, our images come from. And then we may question their validity.

5 OTHER REPRESENTATIONS

The idea of representation applies to other aspects of culture apart from people.

Just as people are represented to us in terms of categories and characteristics, so also one can say that institutions such as the police or ideas such as that of family are represented to us in various ways. This happens in genres especially, but also in other material. As with people, these representations are reinforced across the media – intertextuality at work again.

In fact these other aspects of representation are usually mixed up with the representation of people by type, and end up being discussed anyway. For example, one is likely to talk about how the media portray old age if one talks about how they portray old people. Again, if one teases out the full range of meanings in a genre text, perhaps through structural analysis, then one may also end up talking about the meanings given by representation. For example, in soaps, one kind of structure is provided by the different families and the pattern of relationships within the family. Often one family may be opposed to another, or members of a family will be in patterns of bonding and conflict. But in talking about what that structure is and why it matters in terms of making sense of the story, then one is like to say something about how families are represented – as a shifting pattern of enmities and alliances.

One is getting to the same meanings in media study by different routes.

This broadening of the term 'representation' draws attention to the fact that the media offer us versions of reality. Genres in particular offer us formulaic and fairly predictable versions of people, institutions and aspects of our culture. Genres include material such as news. In this case the representation is largely not fictionalized, and is closer to life than, say, an escapist action movie. Yet even here it is worth noticing that whatever the medium, news is a re-presentation of events, of people, of views. It is still a version of the world.

The word representation can be instructive if it demands that we ask questions such as:

- how is this topic being represented?
- through what particular devices?
- why is it being represented in this way?
- in whose interests is this representation?
- what is really being said about the topic being represented?

In *The Media Studies Book* (1991) Gill Branston talks about the 'assumptions, "common knowledge", common sense, "general" knowledge, widespread beliefs and popular attitudes' which exist in our culture, which we have in our heads and which form a background to representations of people or of institutions. These items could also be said to give us a set of norms (what is right, true, proper, OK) which help us make sense of representations. (The notion for example that heterosexual marriage is OK, but a gay marriage is not.) But the traffic is also two way. It isn't just that general assumptions and beliefs help us measure representations. It is equally true that representations feed back into these beliefs, to confirm or oppose them. This circularity creates dominant meanings, and reinforces the dominant aspects of ideology.

In discussing representation, McQuail (1992) refers to the example of crime reporting with reference to various pieces of research over the last 30 years. He says that 'the media consistently underplay petty, non-violent and white collar offences and emphasize interpersonal, violent, high-status and sexual crime.' So it is that we acquire a notion of categories of crime and categories of criminal.

6 PRESENTERS, PERSONALITIES AND STARS

6.1 The Importance of Personalities

The media rely on 'star characters' as a means of contact with the audience for much of their product. These characters achieve both larger-than-life qualities and also a comfortable familiarity, so that for the television audience they are integrated into their circle of friends. People go to see an actor from a radio series such as *The Archers* open a fête: they go to see the character from the drama, not the actor. People write in to agony aunts as if they were an intimate friend. Readers write to the editor of comic books such as *Superman*, discussing the life and history of the hero as if he were a real person, not a figure of fantasy. Of course this says a great deal about the audience and its needs and perceptions. But the media collaborate in this familiar relationship. They do it of course because it makes for good ratings and sales. People are interested in people. What the audience is encouraged to forget is that they are not dealing with real people, only a symbolic version of a person represented through forms of communication. And even if I now continue to refer to the real person behind the newspaper article or in front of the camera, then still the character that we perceive is different from this real person, it is constructed through the medium.

Such characters are only representations. **They are constructed like other representations (and stereotypes).** I will also concentrate on television as a source of examples, because there is a distinctive quality to the medium. It is essentially iconic – it reproduces the images and sounds that closely resemble real life, whereas the persona of the columnist has to come less directly through language. The difference is significant.

6.2 Stars

'Stars' are actors or actresses who achieve enormous status because of the attractive persona which they project. It is significant that the word usually refers to film and not to television. The international scope of the film market and the previous dominance of the film industry (which after all invented the star system) have contributed to this. Another crucial difference is that film is known to be something made, an edited, fixed story that is not happening as we watch it. But television is quite often live and very often deludes the audience into thinking it is live even when it is actually pre-edited or just a repeat. Again, the viewing conditions are different – cinema-going is a more exotic activity than sitting in one's living room. Even now cinema stories often operate on a scale, within backgrounds that the small screen can neither afford nor do justice to. So it is that film stars also lend themselves more readily to the archetypal and the mythic, though some achieve status for sheer ability and range. Clint Eastwood was well known as Rowdy Yates, a character in a Western television series many years ago. But it took movies to make him a star. Through these he has become a mythic character that quite transcends the parts – a man alone, a character who has as Dirty Harry and the lone rider become a dispenser of something like divine justice beyond the reach of ordinary mortals. The other kind of star – of immense ability and international reputation – is represented by Meryl Streep. She has elevated herself through her sheer ability to convince across an enormous range of parts.

What you need to do in dealing with the SO WHAT question is to consider firstly how the persona of stars is made up, and secondly how it appeals through the medium. Because whatever it is that makes up these screen personalities (and even a journalist as personality or TV star) is something which appeals to the values of the audience and which gives them pleasure. These are powerful points of contact. These bring in the punters. These make the films successful. Whatever value messages the star character communicates, these are the more significant because they are seen by so many people and appear to be so attractive to those people. And the messages communicated through the film as a whole are also likely to be worth examining because they are being reinforced by repetition to audiences in various countries.

The **star persona** is a set of characteristics projected by the star through the roles that they play. It is, if you like, the dominant personality traits which we the audience read into the star's peformance. It is arguable that some of the most successsful stars can be 'read' positively in slightly different ways by different audiences, particularly different genders. For example, much has been written about Madonna, who can be read as sexually attractive by men, and as strong and sexy by women.

The star as cultural myth has much to do with how far the persona is in tune with the desires and beliefs of a culture at a particular time. Eastwood achieves mythic status for a number of interlocking reasons. One is simply that he has been so successful for so long in the entertainment world: famous for being famous. Another is that some of the parts he plays, as in *Pale Rider*, are actually designed to present him as possibly a figure from another world, a mystic hero, justice personified. But a dominant reason is simply that he is in tune with his times, with the dominant ideology. Whatever the current debates about gender roles, the fact is that Eastwood represents a kind of masculine ideal. His characters act decisively, talk sparingly, have a sense of irony, embody individual prowess, dispense justice, and are physically capable. They are single-minded yet humorous, and can display compassion and sentiment. And they do what every bank clerk can only dream of. Myths are about dreams.

Stars have commercial value. They can sell pictures and programmes. If you want to do a deal to make a picture then you need a package. Part of that package must be the star. The star's appeal to the audience can help sell the package and make the deal. And even once the picture is in production, **the star**

becomes a marketing tool. The star image appears on posters, the star appears on TV programmes or at film festivals. Stars boost profits. So it is that even a successful TV anchorman as star can secure a good contract because his appeal makes for good ratings.

So stars are very important to media product, both in terms of commercial profit for the institution, and in terms of attraction for the audience. They are also important to media study because whatever it is that appeals says a lot about ourselves, our beliefs and values.

You pays your money . . .

■ Estimated cost of booking these personalities as panellists on a TV quiz. Compiled by Dave Wood

£1,000–£5,000	£5,000–£10,000	£10,000+
Zoe Ball	Jo Brand	Michael Barrymore
Julia Carling	Nick Hancock	Cilla Black
Emma Forbes	Bob Holness	Jim Davidson
Tony Hawks	Ulrika Jonsson	Angus Deayton
Lee Hurst	Vic Reeves	Chris Evans
Mark Lamarr	Shane Ritchie	Clive James
Helen Lederer	Gaby Roslin	Paul Merton
Des Lynam	Jonathan Ross	Bob Monkhouse
Donna McPhail	Chris Tarrant	Frank Skinner
Jo Whiley	Dale Winton	Anthea Turner

Fig. 5.8 Likely cost of booking given personalities on a TV quiz show

Figure 5.8 indicates that the persona of a star or of a personality really does have a price on it. Ask yourself why there is a difference in the 'price ranges'. A clue to one answer is also to ask yourself which of these names you know or do not know.

6.3 Presenters and Personalities

The same argument applies to the presenters and personalities of television. They are crucial to our understanding of most programmes. They, like stars in film, attract us to the programmes. They also carry meanings within the persona that they project, and bring us to the programme as a whole, with all the other messages and meanings which this may hold.

Many programmes rely on the presenter as a point of contact: chat shows, quiz shows, news, documentaries, current affairs, magazine, consumer advice. The presenter becomes another person in the living room. The personality creates the atmosphere for the show – jolly, serious, knowledgeable, or whatever. **She or he becomes the narrator of the storyline which is unfolded through the programme.** The presenter often explains things to us, introduces people to us. Most of all the presenter has the privilege of talking to the camera. The director (or the editor) makes sure that they do. They are given the screen time to establish their personality and their relationship with the audience. Other people are denied this privilege. There is no justification for this convention, but it is very powerful. News reporters or even chat show hosts hate it if the subject addresses the camera and so the audience. This upsets the mythology of their special relationship with us, the audience. The power of the look, of face-to-face address is considerable. The viewer is directly engaged by the eyes, which is very different from simply watching a vox pop in the street or a conversation in the studio.

So **you should look at what the presenter does for the programme, and how they shape its 'story'**. You should consequently consider the power they have to shape the meaning that we take from that programme, especially as their position is endorsed by addressing the camera (often helped by the invisible autocue). I also suggest that you evaluate the significance of all this by working out what happens when the rules are broken. Why does it matter when the vision mixer makes a mistake and cuts to a side view of the presenter addressing us? Why did it make the papers when the comedian Rod Hull's puppet emu apparently assaulted a chat show host and provoked a flash of temper on air?

Part of the answer lies in the fact that presenters contribute to the 'seamless robe' of television. **They are there to keep things running smoothly, to make sure that we are not aware of the technology which makes it all possible.** It could be said that they are there to stop us asking questions about what we are being told. Even the presenters of quality documentaries are part of this mythology of the effortless power of television. They too can become elevated by the medium. The historian guides us through places, documents, reconstructions, as if a god, and much like the newsreader.

Television also relies on the personality of the characters who front its drama, especially the soaps, *Roseanne* or *EastEnders*. It is arguable whether these well-known people should be regarded as stars or personalities. If there is any distinction to be made it is in respect of awe and intimacy. The stars of film are unattainable and untouchable.

Often their parts are not of our everyday experience. But many of television's fiction stars have roles within our life experience. They are not heroic. They make publicity appearances as they try both to trade off their popularity and yet also enhance it by being seen. These television stars are, by and large, much like the presenters – even though they do not talk to us they create an intimate and familiar relationship between the programme and the audience.

The significance of presenters and stars in the media is partly that they provide a crucial bond between the medium and the audience. They are a commercial element within the product. They pull in the audience. They perform certain functions according to certain rules, which affect how television programmes in particular are understood.

Fig. 5.9 From BBC *Nine O'Clock News* and *Channel 4 News* – Michael Buerk and Zena Badawi

Presenters on television have a good chance of becoming 'personalities' because of the exposure they get. They become familiar. They stand for the programme they front. The two news presenters in the pictures are an example of this. You can ask yourself what they 'say' about the programmes. Do they have a special persona which helps the authority of the news?

7 FACTS AND FICTIONS: REALISM

There is another important dimension to all this media material which comes off the production line. This has to do with the idea of realism.

At least for some of the time, **we judge a great deal of media product in terms of whether or not it is realistic**. This is especially true of story-type product. So it is very common to hear people react to a film with phrases such as 'it wasn't very realistic, was it?' Even the lyrics of popular music can be judged in this way: 'it was just like something that happened to me, it was so truthful'. And we expect newspapers to be about things that have really happened. So it is important to look more closely at 'realism' because it suggests that some media product is more believable, and so we are likely to take it more seriously, and to take on its ideas more readily.

But the first problem that this idea raises is simply, what do we mean by realism? There is a whole set of words which we use in various ways to define realism, without thinking about it. It is these words which we should now look at.

7.1 Definitions

Believable or Credible: what we see or read is something which could have happened. That is to say it resembles the world as we know it. (But remember that there is a lot which we believe to be true, but which we have not really checked out for ourselves – so how violent are the streets of San Francisco?!)

Plausible: what we view or read is at least possible within its own terms of reference. Someone could have acted in the way they did in a given story, or the development of the storyline is basically possible. There is some consistency in the material, even when we know it is basically fiction. So for example, we may find it implausible to be told that the murder has been committed by the long-lost twin brother of the accused hero.

Probable: has much to do with ideas of cause and effect. We tend to talk in terms of whether or not it is probable that one event would follow from another, or whether it is probable that a character would have taken a particular course of action.

Actual/Actuality: the material seems to have an immediate kind of physical reality about it as if it is really happening before us, or even as if we are really there. Often documentary material has the quality of actuality.

Verisimilitude: this word, like 'actuality' suggests that something is true to life. But we also tend to use it when we feel, for example, that people's behaviour has an authentic quality, that it is like life (as we believe it to be).

Truthful: this is an important word because material doesn't have to be entirely believable in a literal way to seem truthful. A story can say something truthful about human behaviour and motivation, even when it is improbable in terms of its situation and background. Many plays, not least Shakespeare's, are fairly improbable in terms of storylines, and certainly in terms of how real their settings are. But they might say something important about the beliefs and values of the characters, which the audience agrees with. These beliefs and values then become the 'truth' that we are talking about.

Some Criteria for Realism

So when we say that something has the quality of realism we could be talking about a number of elements:

- how accurately the background is depicted
- how believable the behaviour of people seems to be
- how probable the storyline is (if we are talking about fiction)
- how true the points made by the material seem to be.

In all this the complication is that realism is all relative. It is relative to our experience. So if we have experienced or even read about something which then appears in a magazine, we may find it more believable than does someone who has not had that experience. It is also relative to the mode of realism.

7.2 Modes of Realism

These refer to **the categories of realism** that we learn and have in our heads when we are making **judgements based on ideas such as realism, truth, believability**. We change the basis of our judgements according to the mode of realism that we think we are dealing with. We do not expect a computer game to be all that realistic; we do not expect it to look as real as film material, nor the situations to be as plausible as those we read about in a newspaper. We do expect a TV documentary to be realistic; we expect it to be more real and believeable than a romantic novel, for example.

7.3 Conventions

We are back to these hidden rules. The fact is that **all these different modes of realism have different rules**. A change in the rules changes what is expected. The particular medium, or mode within the medium, has particular expectations. These expectations are aroused as soon as we start reading, viewing, listening. We expect an autobiography to be different from a novel; a situation comedy to be different from a current affairs programme, and so on. We have prior knowledge about the newspaper medium, we do not expect it to make up stories. We will have read reviews or publicity material about a film and so will know if it is fiction, and even what kind of fiction. *Ghostbusters 2* is not going to be the same in terms of realism as *Apollo 13*. In the case of television, programme title sequences are crucial in letting us know what set of conventions our brains should switch into before the main part of the programme starts.

The rules that we are talking about come across in many different ways. For example, in situation comedy canned laughter is acceptable. In documentary, long shots of someone talking to the camera are acceptable. In radio

Fiction	Documentary
Controlled lighting	Natural lighting
Re-recorded sound	Natural (live) sound
Multiple camera setups	Single camera setup
Actors for characters	Real people
Music	Music infrequently
'Invisible narrative'	Narrator V/O, narrator to camera
Mobile camera	Fixed camera, usually
Studio sets and locations	Actual locations
Editing pace	Shots held/lack of editing pace
Unfolding drama on screen	Interviews on screen
	Captions on screen

Fig. 5.10 Realism – film and television – dominant conventions

journalism, recordings of someone talking through a poor telephone link line are acceptable. In film fiction, sudden bursts of romantic music are acceptable – you can easily extend these examples. But generally we don't mix these sets of rules. So once we have locked into a particular set of rules for a particular kind of realism then we have set up particular standards for and expectations of the quality of realism in the product.

7.4 Realism, Narrative, Ideology and Genre

John Fiske says in *Television Culture* (1987) that 'realism does not just reproduce reality, it makes sense of it'. He is drawing attention to the fact that if one aspect of realism is about content – what looks and sounds real – another aspect is about form – how things are put over. This matter of form also refers back to the last section about conventions.

In this respect it is arguable that realism has a lot to do with narrative – how one tells the story, or rather how one uses the medium or form in order to cause people to construct a story in their heads. There is a 'typical' way of telling a story in our culture in various media, without arguing too much about the shuffling around of various cards in the deck of conventions, This usual way of making a narrative can, I suggest, be called mainstream narrative, or a classic realist text. The two phrases lead one to the same thing, albeit from slightly different perspectives.

One could go further and argue that the conventions of narrative are to an extent exactly the same as the conventions of realism – they are two aspects of the same devices of form. The point here is that mainstream narrative in any form strives to divert attention from itself. Storytellers don't want you to realise that they are telling the story, and that it is really just a collection of devices for making up the story, then it is a short jump to the idea of realism. It seems real because there is apparently nothing between you the reader and viewer, and the stuff of the story – you are there, you are in the story, it is happening before your eyes. This 'trickery' abounds in every example of the media. For instance, because we are supposed to be watching a news interview as if it is really happening, the producer puts in 'noddies' of the interviewer apparently asking the questions – though these were not shot at the time, and are not needed for any practical purpose. We could just as well hear the questions as voice-over. But we are put in the classic third-person narrative position that seems so realistic to us, by being made able to apparently see the interviewer as well as the interviewee.

At the same time, realism is tied in with ideology. That is to say, stories have meanings about beliefs and values in particular. These beliefs and values are the stuff of ideology. There are dominant values in the generality of popular texts (the dominant ideology). Realism helps make them dominant because the idea of what is real is tied in with the idea of what is true; what we think is true, we tend to believe. Many media critics have also argued that realism helps make dominant discourse dominant. So if you get sucked into a romantic story on television, or an adventure novel because it all seems very believeable at the time, then you are also absorbing dominant discourses about masculinity or femininity. It is likely that you will be absorbing and reinforcing covert values about what it means to be male and female. The novel will be about physical prowess and risk for the male, and about relationships for the female.

Fiske points out that there is an argument here for saying that 'all popular culture inevitably serves the interests of the dominant ideology', that it 'provides the common ground between producers and audience-seen-as-consumers', that popular media material positions 'the viewer as a subject of and in the dominant ideology so effectively that any radicalism of the content is necessarily defused by the conventionality of the form'. So if a sitcom such as *Men Behaving Badly* raises questions about masculinity in the content of its stories, it completely undermines serious questioning by the way (form) that it makes a joke of everything and carries you along through conventional storytelling.

This example and argument also makes a link with genre and demonstrates that in media studies it is difficult to talk about one major concept without referring to another. My argument has suggested that certain discourses are more likely to appear in certain genre material than in others. For example – the privileging of female discourse in romance or in soaps. But it is also true that certain modes and conventions of realism are linked to genres, if not in an absolute or rigid sense. For example degrees of naturalism and authenticity attach themselves to soaps, but horror stories do not assume naturalism, indeed they allow for a degree of fantasy – and we adjust our expectations accordingly.

7.5 Sources of Realism

It is important to remind ourselves that all our views about what is real or truthful depend on a number of kinds of experience. One is **cultural experience.** We draw on the years of learning throughout our lives about what our culture sees as real. For example, we have learned a language of visual imagery and so believe that a larger object concealing a smaller one in a picture is closer to us than the smaller one. But this is just a convention of the code of visual communication. It is just another one of these sets of rules. Another culture which has not learned the rules this way would say, what a silly picture (painting, photograph or film) – the creator has put one thing in the way of another – or – the bigger object must be more important than the smaller one. And of course all the sets of rules for the various modes of realism are learned through our upbringing in our culture.

Another experience is that of our real life. That is to say we may judge what is or is not real on the basis of what we have seen, done, felt. In particular, we may judge realism in terms of probability from our life's experience. From life we know something about cause and effect, about likely human behaviour. So then from that experience we can judge whether or not what happens in the media is probable.

But yet another experience is **that of the media themselves.** This is also woven into our reality. If we have seen part of a documentary about American Indians, and then watch a TV drama set among Indians in New Mexico, we will judge the realism of that drama partly in the light of the other piece of media material. The point is that we may never have been to New Mexico in our lives. So we will base our judgement on this second-hand media experience which someone else has created.

At the same time we should remind ourselves that the credibility of the course also matters. So it is possible to argue that while there is less output of documentary on television, its credibility and reference to reality out there weighs more in the balance than the majority of fictional material. That average of 20 hours per week on terrestial channels, which represents 2 hours per year that the average documentary maker produces, has a particular 'weight'.

7.6 Realism, Production and Genres

Realism affects how we relate to media material, especially fictions. It is important to understand it because it affects the credibility of messages in the material. And for both those reasons the media producers find it important to maintain and promote the various modes of realism and their rules. It is convenient to package material into kinds of realism just as it is pre-packaged into genres. The producers want the audience to feel comfortable with their product, to know where they stand. They want the relatively realist modes such as documentary, or media such as newspapers to have credibility.

It is especially important in the case of television that viewers are able to distinguish one programme from another in the endless stream of material. If

you pick up books from shelves labelled 'fiction' or 'travel', then you have a pretty good idea about their kinds of realism before you start. But television is like an endless procession of open books in different modes, without the shelf labels. This is where the titles sequences and programme previews come in, as well as the cues in the programmes themselves.

It is perhaps no accident that some of the most hot-tempered public debates about the media revolve around television in particular. The arguments, while

Fig. 5.11 Stills from *Middlemarch* (opposite) and *EastEnders* (above)

The pictures from *Middlemarch* and *EastEnders* represent two qualities of realism. The first is about authenticity, with attention to historical detail and accuracy, so that one feels that one is 'really there', even though we are only too well aware that it is a story taken from a book. In the second case it is about naturalism as well as authenticity. And we know that there really are pubs with interiors like this one. Perhaps the realism of the soap is bound up with its serialization on the screen – it becomes part of our life. Perhaps the realism comes from the working-class subject matter.

sometimes seeming to be about programme content, are actually as much about programme treatment. In other words, the debates about bias in news on television are also debates about how items are handled, about the fact that it is generally believed that people believe what they see on the news. Similarly, the debates about certain dramas are often about how conventions of realism are used. *Our Friends in the North* was a British series based on real political events, including corruption. Drama documentary productions such as this example are by definition a blurring of the lines between fact and fiction. The argument about this drama had a particular edge because it looked pretty real and because parts of the storyline had some connection with known events in real life.

A strong example of the relationship between realism and product, especially genre product, is that of TV soaps. Everyone 'knows' that they fall generally within the mode of fiction. But within this broad category there is still a scale. They are in a different part of the scale, we tend to assume, from a horror movie. On this scale we expect soaps, especially British soaps, to have locations or characters from life. Kilborn (1992) says that they 'seek to create the illusion of a reality', they have 'a sense of lived experience'. This quality of realism is in the authentic detail of sets or of dress. It is in the extended time span that is a luxury of soaps running over months and years, the possibility of matching real time. It is in the ideological dimensions of realism, where soaps tend to reflect attitudes and values, even their shifts over a period of time. The very longevity of some soaps makes them part of our life's experience. This in itself causes them to become part of our reality. I think that we accept a rather different kind of realism in soaps from other cultures simply because they are from other cultures, and we can't have the same kind of life reference. So British soaps are an example of a kind of look across from kind of product to kind of realism.

7.6 News and Fiction

The truth of **the matter is that there is no absolute reality or truth in the media.** We may like to think there is; we may find it convenient to have sets of rules to define relative kinds of reality and truth. But look at the facts. A newspaper such as *The Independent* reads as somehow more realistic than *The Mirror*. The former has more hard news stories and a less dramatized style than the latter. And yet one cannot say that because *The Mirror* may prefer a majority of human-interest items (perhaps a film star's divorce) to an item about what is happening in the House of Lords, that the divorce item is not actually true – not real. Similarly, television news, while not being simply untruthful, has qualities of drama in the way that it selects some exciting stories or makes excitement out of something like a kidnap story. And it certainly is not simply THE TRUTH. If it could achieve this then we could make do with one news programme only. So the lines between one kind of realism and another may be more blurred than we think.

7.7 Packaging

We are socialized into believing that the divisions are sharper than they are through the various ways of packaging the product. It is packaged through generic labels (thriller or soap); or through treatment (fiction or documentary); or through audience targeting (for women or for sports fans); or through structuring (a page of adverts followed by a feature article or a soap followed by a quiz followed by a sitcom). All these various ways of packaging the media product aim to get the product and the audience together to achieve maximum sales, maximum ratings or readership, maximum profits.

From one point of view this makes the process of communication as smooth, as seamless, as effective as possible. What it also does is to project the messages very effectively. So once more we will have to return to the issue of what these messages are, whether or not they are intentional or covert. To this extent the product is not just the book, the record or the programme, it is also the message. If there is a production line, a process by which material is produced repetitiously, efficiently, in quantity, then it is also a message production line. It is **a production line of values; it is a reinforcement of values.**

8 NARRATIVE

The media tell stories. These stories are not just fiction, they are about factual material as well. Newspapers talk about 'the story' when they refer to a piece of news. Preferably we should use the word 'narrative'. **The media are full of kinds of narrative,** if only because all these media have to unfold their material in sequence. Just as a radio drama will introduce characters and situations at the beginning, a magazine article introduces its subject and theme at the beginning. Just as the drama goes on to unfold a story with themes to bring out, the article unfolds its information and the points it wants to make. **There is narrative whether one is talking about fact or fiction.**

And even if we are just talking about fiction it is worth remembering how much of this there is in the media. Magazines and even Sunday newspapers carry short stories. Comics are full of stories. Many popular songs are based on a story. Video games are stories that you take part in, and originate in fictions (see how products like *Judge Dredd* are actually marketed as the film, the video and the video game, as well as the comic).

We have already talked about the presenter as a kind of narrator for television product. But let's remember that the narrative is organized before anything appears on screen. The narrative of a television documentary is organized through a script, through directors' and editors' decisions. The 'storyline' for a newspaper is arranged by the subs and the editor deciding what goes on what page. So the idea of narrative draws attention to the way in which material is organized for consumption by the audience. It can make the material and its ideas more digestible. At **the heart of narrative is meaning** – what the

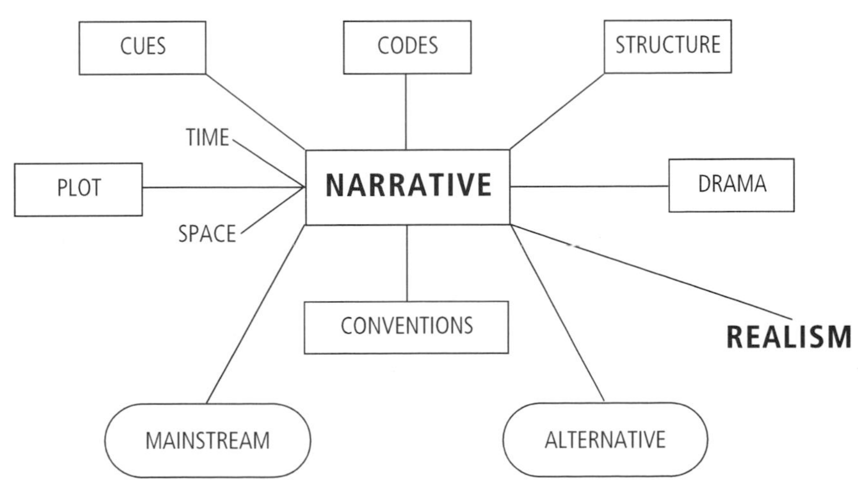

Fig. 5.12 Key concepts – Narrative

story is about. How we understand what the story is about depends on our ability to decode the narrative. This is what we are now going to look at.

8.1 Narrative, Space and Time

One crucial thing which narrative does is to shape the material in terms of space and time. That is to say, **it defines where things take place, when they take place, how quickly they take place**. Even when television is broadcast live it can still pull off this narrative manipulation. In terms of space and place, you can be watching a sports programme being unfolded from a studio in London. The narrator is based there. But this narrator/presenter can project us to other places where sporting events are taking place. We can be sent to Las Vegas to watch a boxing match. At this boxing match the narrative may be shifted in time. For instance, we have the replays of crucial points of the action. These have the effect of distorting real time and events. Often the replays are used between rounds, so that we lose a sense of the stop/go action, and are given an illusion of continuous activity in the ring. The replay is also a distortion because it is a flashback. Fiction on screen is very prone to use this device, to compress or lengthen screen time. Sergio Leone, director of Westerns, developed a trick of spinning out real time on screen when there was a moment of tension between characters. He used very long-held reaction shots around faces before there was an explosion of real action, in order to heighten tension by making the viewer wait for it. So narrative, especially that of film and television, has an immense ability to manipulate our awareness of time and place. This is another kind of manipulation of meaning.

8.2 Narrative Modes

We can recognize at least two modes of narrative which need to be structured. One is **the narrative of events** – things that happen and the order in which they happen. The other mode is **the narrative of drama**, which is more to do with character and relationships. In this case the narrative structure is marked by things like dramatic tension and crisis in relationships. If the hero shoots the enemy agent, dives into the lake and triggers the remote-control device which will destroy the . . . whatever it is, then this is one kind of order of events. If the heroine has a tense argument with the hero and decides that he was never her type and that she is going to leave, then nothing has happened in terms of events, but a lot has happened dramatically – there has been conflict, a change of relationship, and the story is about to change direction.

8.3 Narrative Structures

Most narratives are structured on the basis of an opening and a closure. In principle one does not have to do this – you can just cut into the storyline. But in fact often there are devices used which say, this is the beginning, this is the end. A newspaper article is opened by its headline and closed by some verbal exit line and by a graphic marker such as a rule line. Television programmes are opened by titles and closed by credits. This kind of structuring is designed to help us make sense of the material. Television especially needs this, so that we know where one programme begins and another ends. It needs it especially because it also has a larger kind of structure which runs over the whole day. It opens with breakfast time and closes with the late film and announcer's finale – or whatever.

This sense of structure extends through the whole piece of media material. **It organizes the order of its 'telling'. It organizes the understanding of the audience.** *Top of the Pops*, showcasing chart hits and hopefuls, has a very rigid structure. It requires alternation between the presenter DJs and the bands on stage. There are three ritual run-downs on the charts. There is always a number (with video) to play out at the end. The very music it plays also has structure. One example would be chorus–verse–chorus–middle eight–verse–chorus. Magazines and newspapers have a pretty regular overall narrative structure. This is what helps to sell them to the readers – the comfort of finding items in their familiar places – the poster pull-out in the middle, the sport always at the back, and so on. Yet again, the individual articles will have their own structure. For longer newspaper articles, the paragraphs and subheads mark this out, as information about the event is added, and the angle is pursued.

There are various structures which shape fiction. One kind of circular structure shows the dramatic event (a murder?) which is the impetus to the story at the beginning. Then the main part of the narrative is a flashback which shows events leading up to the murder. By the end we have reached once more the point in time where the murder took place. Then the murderer is discovered. In the most conventional narrative one is introduced to the protagonists (heroes and villains/lead characters); and to the background; then to a problem or some sort of conflict which has to be sorted out; then taken through a set of events in which the structure is about 'will it or won't it get sorted'; finally, just when everything seems impossible, there is a grand finale in which everyone gets their just desserts.

You should try analysing the structure of various examples of narrative to see how they are organized. Remember that the particular point is to hold your attention, to arouse your emotions, to give you this sort of pleasure in viewing or reading – ultimately so that you come back to buy more.

And don't forget the narrative of non-fiction forms. News is prone to create drama. We accept that fiction generates high spots of dramatic tension. We are used to experiencing what we call 'cliffhangers'. Serials in comics deliberately engineer these so that you will buy the next one to find out what happens. News does the same thing. It promises pictures of the disaster later in the programme. It invites you to watch the next episode on a major story later in the evening in the next news broadcast.

The notion of structure in narrative has led to various attempts to find something universal in the organization of all stories. Fiske discusses this at some length in *Television Culture* and refers to the work of structuralists such as Propp who sought to analyse folk tales down to such univeral elements or rules – the ultimate structuralism. It doesn't work, but the attempt is valuable.

For example, such analysis leads to a recognition of **binary oppositions** referred to elsewhere in this book. And the identification of an oppositional structure leads to a realization that it operates on a literal and a symbolic level in stories. In this case Fiske points out that 'the struggle between the hero and the villain is a metaphorical transformation of that between the forces of order and those of disorder, good and evil, culture or nature. Such a struggle is fundamental to all societies'. In other words, opposing characters stand for opposing ideas. The narrative has a structure which works on the level of character and action, and on the level of concepts, discourses, myths.

Furthermore, such a structure relates to notions of resolution and contradiction. That is to say, to an extent, stories work to resolve the opposition. The story of *Othello* is partly about an opposition between duty to parent and duty to husband; Desdemona resolves this by saying that the act of marriage means that the weight of her duty now shifts towards her husband Othello. But, like many stories, this rather avoids other contradictions which are never resolved. If she had obeyed her duty to her father, she would never have sneaked off to get married secretly in the first place. Anyway there is a related contradiction which is that the Venetians see Othello as a capable general, but also as a Moor. And nice young Venetian girls are not supposed to marry Moors. This opposition based on race is never sorted out. It is in a sense avoided by having Othello murder Desdemona out of jealousy. It would have been far more interesting to have a story in which Desdemona lived on and had children, How would that have been sorted out?! Such contradictions in narratives, in our ideology, are quite frequent. The same story contains one classic opposition which is about wanting women to be sexy and exciting and pure and loyal at the same time. It is suggested that Desdemona is both lustful to want to marry Othello yet also pure and innocent because of her class background and because she is a victim. Her murder also avoids dealing with this one.

8.4 Narrative Cues

These cues direct our understanding of the unfolding story. They may be verbal or visual. They tell us things like: this is a villain, this is a time shift, the story is now moving to another place, something dreadful is about to happen, and so on. To take more obvious examples, the villain is cued by a close-up on an unshaven face, the time shift is cued by a pull-in on the character's eyes as they remember something that happened before, the place shift is signalled by 'and now we go over to our correspondent in Beirut', the anticipation is cued by a cutaway shot to a hand reaching for a gun. You can make your own list of such cues. But always ask why they are there. Consider what they add to the meaning of the story.

8.5 Narrative and Conventions

Narrative is as much bound by **rules regarding the way it is handled** as are other examples of media content and treatment. We have already noted that narrative cues are, for example, conventions themselves in terms of how they are used and what they mean. They are rules for organizing the narrative. And as with genre or realism, it is important that those who make the communication share the same rules with those who receive and decode it. This creates understanding. Unconventional narrative or unconventional use of the rules can be upsetting, confusing and alienating. This is not to say that the rules cannot be bent, or combined in different ways. It is certainly not to say that people should not experiment with different ways of narration. For example film makers have tried using black screen to mark a separation between one episode and another. Some have tried killing off the hero half way and introducing another. Some have jumbled up the order of events, and used flashbacks within flashbacks. It is useful for a media student to watch this kind of film because it really proves that the rules for constructing narrative do exist. But still it also proves that the audience has to learn the rules or make sense of new ones if it is to create meaning from the material.

Conventions of narrative may overlap with those of genre or of realism. It is like having one lever which can move different bits of machinery. For instance, you might have a close shot of a hand spinning the chamber of a Colt 45 handgun. The fact that it is a Colt handgun and therefore an icon of the Western makes it a conventional element of the Western genre. This in turn will incline us to drop mentally into a certain mode of realism – straight fiction. But the image also sets up the narrative. We would expect a scene to follow in which the gun is used.

So once more, in terms of the creation of meaning, of the exchange of meaning between the media and its audience, we must not underrate the importance of conventions in making this possible.

8.6 Narrative and the Reader/Spectator

Narrative has the power to place us in a relationship to the story. One fundamental relationship is that of the objective or subjective position. In the first case we are standing outside the action, in the second we may be drawn into it. One is first-person storytelling, the other is third person. Visual and fiction material provide the strongest examples. It is possible that the two kinds of positioning will happen even within one scene. If the heroine is hanging from a cliff top and we are simply viewing this from a distance then this is objective narrative. But if the camera switches to a view down the cliff as if we were in her eyes and her head, then this is subjective narrative. Print media do the same thing when they use the 'I' form to address the reader directly, or when they simply describe events in terms of 'it happened'. Factual material can use the same device – documentaries may drop in subjective shots to give the experience more impact – perhaps in a travel programme. Newspapers tend to stick to third-person narrative because they are trying to achieve objectivity. But magazines may well address the reader directly in order to achieve intimacy and credibility.

This kind of positioning is calculated to affect the emotional impact of what

Fig. 5.13 Visual narrative: a page from *Cyclops & Phoenix*

Comics provide interesting examples of narrative because they mix their own conventions with those of film. The page from the comic is the 'story of a conversation', which is itself about modern and Victorian ideas on evolution. The visual narrative of the frames alternates close and wide shots, changing implicit 'camera angles'. You might look at other aspects of the visualization, which tells us about ideas such as that of male power.

is being told, to affect its believability. To this extent these devices of narrative also relate to realism. The way in which we relate to the narrative affects how we react to it and affects how we accept the meanings or messages. Third person in newspapers provides credibility – so we go along with the meanings.

Narrative can also place us in **apparently impossible positions** in relation to action. We could have watched our heroine on the cliff top as if from some point in the air looking at the cliff. We accept this godlike position because as with everything else we have learned to accept it from previous viewing. We have learned what are yet more conventions of storytelling.

Then there is **the privileged spectator position**, when we can see or know things that the protagonists do not. This should remind you in terms of image analysis that it is the power of the camera and of those who direct it which puts us in a certain position. So we may be enabled to see a crucial document on a table which someone else within the story cannot see. At the same time, this device also draws us further into the drama. It is a sort of participation in what is going on. You can see the same thing happening in a quiz show when the presenter not only welcomes us personally (though he cannot see us) but may show us the prizes, which the quiz shows players cannot see.

There is the positioning with regard to the use of a narrator. Detective movies have a convention of voice-over in which the hero confides in us about what is happening. The newspaper reporter acts as narrator when she gives us an eyewitness acount of some event. The first one is more intimate than the second. But both create a relationship between us the audience and the story which is being narrated.

So these devices of narration can put us literally in a certain place, or they can put us psychologically in a relationship to what is being narrated. This relationship is basically about more or less involvement in what is supposed to be going on. The degree of involvement affects our acceptance of everything on the page or on the screen. This in turn affects our acceptance of the messages and meanings within the material. We are not only seduced into thinking this is believable, for example, but we may also be seduced into accepting meanings like – it is OK for people to exact revenge (in a story), or it is OK for people to do stupid things in public if there are prizes at the end of it (in a quiz show).

8.7 Mainstream Narrative

In all media there is a 'generally accepted' way of making narrative, as described in the previous sections. The conventions of narrative mean that we expect a kind of story unfolding which helps us know where we are, who we are dealing with, and to believe that one event or action follows plausibly from another. This is the narrative of the mainstream.

There is quite an array of devices available in mainstream narrative, but one thing they all do is to 'get you into the story', to the point where the reader/viewer forgets that it is just a book or a film – the **narrative becomes invisible.**

The way that the creator(s) manufacture this invisibility does have something to do with the particular medium, words or pictures. But in fact there are qualities of mainstream narrative in any media whether they are achieved by camera or by language – for example, a sense of continuity, linking time and place. Similarly, one would expect a story involving physical or psychological conflict. One would expect some sort of resolution to the conflict or problems of the characters. This last point is known as **narrative closure.**

Realism is so much a part of narrative that another phrase which will do as well as mainstream narrative is – **the classic realist text.** As Gianetti says in *Understanding Movies* (1993), 'the classical paradigm emphasises dramatic unity, plausible motivations, and coherence of its constituent parts'. We are back to the invisibility of the hand of the creator, and to qualities of realism which enable us to accept the text on its own terms. In *The Cinema Book* (1985) Kuhn lists features of a classic realist text or narrative as follows.

1. Linearity of cause and effect within an overall trajectory of enigma resolution
2. A high degree of narrative closure
3. A fictional world governed by spatial and temporal verisimilitude
4. Centrality of the narrative agency of psychologically rounded characters.

These kinds of stories are all wrapped up, they seem to make perfect sense, they seem to belong to a real world, they depend on what happens to characters and what they do. Textual analysis reveals that it isn't that simple.

8.8 Alternative Narrative

One could say that **narrative which is not mainstream is in some measure alternative.** The term suggests that at least some aspect of the storytelling is not within the parameters which we are used to. Of course this creates a problem because it depends on what we are used to. Some might say that Dennis Potter's *Lipstick on Your Collar* (drama for TV) is alternative because it mixes up songs with an ordinary story, dream time with real time. But this attends only to realism as an aspect of narrative. In other respects, the structure, the use of cues and the sense of place and time are all perfectly conventional. In any case it does not make sense that anything which really breaks the rules would appear on mainstream television.

To this extent alternative narratives in any medium are likely to be experimental, on the margins and to appeal only to a limited audience. This is not to denigrate them. Experimental novels – the work of Samuel Beckett perhaps – are praised and enjoyed, though they are outside what is called popular culture.

Perhaps what is most interesting is work which falls between extremes, or is subversive in some way. For example, there are games on CD-ROMs or printed books where the narrative is not fixed in structure and development, but is controlled to some extent by the reader/viewer, who makes choices about how the story develops. There are films like *Koyaanisqatsi* which is really a montage of images with sound and music, and doesn't have character development or plot resolution.

The existence of the alternative in the media, whether it is to do with narrative or audience or mode of production is very important. It is about experimentation and development and it actually defines what mainstream is, and questions that.

8.9 Diegesis

Is a useful term which distinguishes content from form where narrative is concerned. **It describes what is supposed to be in the story and part of it, as opposed to what is outside it, and is to to do with the way the story is told.** If the hero starts playing romantic music in a radio drama then that is diegetic. But if the director puts in romantic music as part of the backing to create a mood for the drama, then that is non-diegetic.

The idea of diegesis also connects with conventions and with realism. Take the example of voice-over. Documentaries can have a voice-over film which was recorded on the spot at the time (diegetic); or voice-over which is dubbed on the film later (non-diegetic). The first device seems a shade more actual than the second, though both examples are within the collection of conventions which help define a factual, realistic mode of film/television. But then again, voice-overs can be used in fiction films – the private eye/hero talking on the soundtrack. The hero is part of the film story, so this voice-over is diegetic. But because it is associated with a realist mode of television which quite often uses it in a non-diegetic, authoritative manner, so the audience finds the device of the protagonist one which enhances the realism of the fiction.

So the idea of diegesis helps one concentrate on ways in which the narrative is shaped and constructed, how we relate to what we think of as the 'story'.

REVIEW

You should have learned the following things from this chapter about media product.

1 REPETITION

- the main elements of media product and the way that these are treated are often repeated.
- this kind of content and treatment has been made 'natural' for the audience by years of exposure to it.
- these elements are nevertheless often attractive and sell well.
- the main consequence is that the messages and meanings in the product are also repeated. They have ideological significance even if they are not intentional.

2 GENRES

2.1 A lot of media product, especially story fiction falls under this heading.

2.2 Genres are built from combinations of key elements: protagonists, stock characters, plots and stock situations, icons, backgrounds and decor, themes.

2.3 These elements add up to a formula shared by producers and by audiences.

2.4 Genre material is quickly recognizable through previous exposure, and is attractive to the audience.

2.5 Genres work on the audience and give pleasure because they are predictable.
2.6 Genres display seriality in repeated story types.
2.7 The content, treatment and messages of genres are reinforced through being repeated.
2.8 Genres not only work to a formula, they also have their own codes and conventions governing what we expect to see and read and how it will be handled.
2.9 Genres give pleasure to the audience and so sell well. This means that they are also profitable to media industries and so tend to be repeated, with variations on the formula.
2.10 Genres create elements of myth in their stories. This is attractive because it represents certain basic beliefs and aspirations in our culture.
2.11 Genres exemplify intertextuality, in which one example is understood by reference to all others.
2.12 We may look at examples of genre to see what the formulae are and what meanings, beliefs and values are carried along with the formulae.

3 REPRESENTATIONS (STEREOTYPES)

- The media construct various representations of social groups by building certain types of people.

3.1 These types may be characterized in terms of types, stereotypes and archetypes.
3.2 These types are built up from repeated elements such as appearance and behaviour.
3.3 These elements carry meanings about character, relationship, and about how we are meant to view and value the types.
3.4 What is not represented is as important as what is shown.

4 REPRESENTATION, CULTURE AND MEANING

- these representations of people say a lot about our culture and our beliefs. They may represent our values; they may reinforce them.

5 OTHER REPRESENTATIONS

- one can talk about other story elements such as organizations or ideas such as old age also being represented.

6 PRESENTERS, PERSONALITIES AND STARS

6.1 Personalities act as a point of contact between media material and the audience. They are used to interpret that material for us. They are also 'constructed' like fictional characters in order to be attractive to the audience.
6.2 Stars are also attractive personalities. The term is usually used with relation to film. Stars have a persona and an appeal which is part of the film product that is sold to us.
6.3 Presenters appear in television in particular. They often act as a kind of narrator to a given type of programme, telling us what it means, perhaps preventing us from making up our own minds freely.

7 FACTS AND FICTIONS: REALISM

7.1 There are a variety of words which we may use to define realism in factual or fictional material: believable, plausible, actual, verisimilitude, truthful, probable.

7.2 There are different modes or categories of realism, for which we have different expectations in terms of how the material is handled.
7.3 These modes are all based on different conventions.
7.4 Realism can be seen as a function of narrative. It can also be seen as being defined through ideology – what we believe is real to us.
7.5 There are various sources for our ideas about realism, notably our general cultural experience, our personal experience of life, our second-hand experience via the media.
7.6 Media producers prefer to package material within one mode of realism or another, so that it is believable within its own terms.
7.7 No mode of realism, whether news or fiction, is absolutely real or truthful.
7.8 Packaging of material in terms of realism is matched by packaging in terms of genres and other categories.

8 NARRATIVE

- This is about the story structure and storyline of all media material.

8.1 Narrative defines the place and time in which things are supposed to happen.
8.2 There are two main modes/aspects of narrative – story in terms of events, story in terms of drama.
8.3 Narratives have various structures, or ways of organizing the sequence of storytelling and what the story means.
8.4 Narrative cues are signs within the storytelling which indicate things like where events are meant to be placed, or shifts of time.
8.5 Narrative also has sets of rules or conventions which producers use in terms of how they tell a drama or a documentary story, for instance. These rules are sometimes also the rules of realism or of genre. These same rules work in more than one way.
8.6 Narrative also includes ways of relating the reader/viewer to the material about which a story is unfolded. Two main 'positions', or kinds of relationship, are called subjective and objective.
8.7 Mainstream narrative is that form whose conventions dominate storytelling in our media, and which is characterized by the quality of making its own devices invisible.
8.8 Alternative narrative is that which 'breaks' the rules.
8.9 Diegesis describes what is in the story (content), as opposed to what is outside it (the form of storytelling).

Activity (5)

This activity is concerned with key aspects of media product – realism, narrative, culture and representation in particular. The emphasis that you put on any one of these is up to you, but you won't be able to avoid dealing with them and therefore creating meanings about them.

The activity also asks you to work in different media. This will help make the point that, while the medium is not all of the message, it certainly makes a difference as to how any subject is presented.

The activity is not an end itself. It is not about just reproducing an authentic example of product and using a few conventions. Even if you only do one of the activities, you will still learn something about how the four concepts can be built into and drawn out of media product.

This activity is in documentary mode, is about where you live, and can have any working title you like – *Our Place* or *My Home Town* perhaps.

The idea is to take a look at your street, your area, your community as if you are a stranger to the place. You don't have to deal in lots of facts and figures or borrow things from sources like the tourist information service (if that is relevant). What you should do is to create a picture of what goes on, of the look of the place, of local characters, of trade and business, of routines. It is meant to be interesting – the sort of thing that a commissioning editor really would want.

WRITE AN 800 WORD ARTICLE as for a newspaper supplement/magazine
SCRIPT A 10 MINUTE PIECE FOR RADIO as for a local radio station
SCRIPT A 5 MINUTE PIECE FOR TELEVISION as for a series of documentary shorts

Obviously it would be good if you could carry things through to recording, and to desktop publishing the article.

In the first place, you can look back at your work and reflect on things like how you have represented people and how you have constructed your narrative.

If you have created all three pieces than you can also reflect on how far the medium makes them different from one another, how it has affected your creation. But are there other factors which have caused you to handle the pieces differently?

Fig. 6 Judge Dredd

6

Special Cases – News: Influence – Advertising: Persuasion

Some Are More Equal Than Others

1 MAKING SENSE OF NEWS

This chapter supplements what was said in the previous one about news as genre. It introduces some more concepts which help explain how meanings are put across in news material. **The reason for giving news such a high profile in Media Studies is that it is a prime source of information about the world** from its geography to its politics. Most people trust the news machine and what it tells us. Often it is endowed with qualities of neutrality and authority which in fact it has not got, and could not reasonably be expected to have. So the ideas which follow will help to demythologize news. What best puts it in perspective as another piece of media communication is the fact that news material is bought and sold every day just like any other product. This leads one straight in to the matter of where news comes from.

1.1 News Gathering

The term news gathering is commonly used to describe the first stage of the manufacture of news. It implies that news is waiting to be gathered in like fruit, and sorted and packed for the audience. But **news is not something complete and fully formed – it is created.** It is not even 'gathered' by the reporter in many cases, as is popularly supposed. A great deal of material comes through agencies such as Associated Press in the case of the press, and Visnews or UPITN in the case of television. The material is paid for. Similarly, the television news operations across Europe have a link up every morning to buy and sell news items. Even where news is collected by reporters it is done in a very routine way for the most part, going to regular sources, using press officers and their press conferences, which front for many organizations, not least the Government. In any case, the news item is not just information from the agency sources. **News is constructed just like any communication.**

1 .2 Agenda Setting

The news organizations set up an agenda of topics which form the news. Once more this opposes the idea that news is somehow a collection of truthful events and facts from 'out there'. The editors **choose the news, and in so choosing also choose an agenda of items which become our view of what is important in the world** that day or that week. Editors decide what their lead items are. In broadcasting they have meetings to decide what their running order of items will be. **Items are selected out and selected in.**

1.3 News Values

News values are concerned with topics that the newsmakers value as being newsworthy, and with ways of presenting those topics.

General Values

Negativity: in general the news machine values the dramatic impact of bad news. Bad news is good news. Events involving a stock market slump or a crash with deaths are rated above a steady market or excellent safety figures.

Closeness to home: news that is closest to the culture and geography of the newsmakers is valued most. So a French yacht that sinks in the Channel may not rate a mention, but if the boat is English it will likely appear.

Recency: recent events are valued above distant ones – hence the competition among newspeople to get a scoop or to break a story first. This value is well projected on public consciousness: people believe that all the news is up to the minute. This is ironic because in fact it is often only major stories that are recent. Smaller items may well be two or more days old. And this value is inconsistent with another one.

Currency: if a story has already been on the news agenda then further details on it are considered valuable, mostly because the audience already 'knows about it'. So stories which run on over days and weeks are not strictly new at all.

Continuity: value is placed on items which are obviously going to have some continuity when the original story breaks. It is attractive to deal with some event like riots or a war, because these are likely to turn into a drama that will run for some time.

Simplicity: items which can be dealt with simply are preferred to those which may be complicated to explain. Particularly, the popular press will prefer a straight story about some act of terrorism to a difficult one about balance of payments problems.

Personality: stories that centre on a personality, preferably a public figure, or which can be developed round a person, are valued above many others because they automatically lend themselves to what is called the human interest angle.

All these general values mean that there are qualities of potential stories which cause them to be chosen above others. To this extent there is bias built in to the newsmaking process. This selective approach to encoding communication is emphasized by other kinds of values.

Content Values

Certain topics will be valued and therefore chosen in preference to others. Examples are stories about disasters, stars, the royal family, authority figures. You can add to this list for yourself.

Treatment Values

These values refer to **what is valued about the treatment of the message,** the handling of the story. Stories which lend themselves to certain kinds of treatment may be preferred above others. Stories may be deliberately handled in terms of these values, even where this does not do justice to the complexity of what has happened.

Stories with **pictures** are valued. **Dramatization** of stories is valued as a way of handling the material. **Conflict** is valued: stories may be told in these terms even when the truth is not simply about A versus B. Stories that can be treated in terms of **human interest** are valued. You will have noticed that, for example, disaster stories are often handled in this way not least because the bare facts soon run out, and interviews with victims and relatives attract the audience. **Actuality** is valued – the news people will put a reporter on the spot even when the spot is very boring. Pictures of a reporter outside a featureless building saying that nothing much has happened so far are quite common.

1.4 Angles

The angle of a news story refers to the particular kind of **treatment or theme which is to be privileged.** Editors talk frequently about the human interest angle for example. They mean that they want the event to be dealt with in terms of the people involved rather than just the facts. **The idea of angle contradicts the notion of neutrality** which the news machine also likes to project as being valued.

1.5 Code

The idea of code has already been dealt with in Chapter 3. It is enough to say that news has its own way of communicating, its own signifiers which we have learned to make sense of. For instance, the live link to the place where the story is based is part of that secondary code. It signifies the authenticity of the item, and is used even when it would be as cheap and informative to have someone in the studio telling us the same thing.

1.6 Conventions

Conventions are unwritten rules about what may be in a newspaper or how it will be handled. This links them closely with news values.

There are also conventions about how the whole story of the news programme is put together and handled. It is a convention that the newsreader acts as link or storyteller. It is a convention that background pictures are put up behind the newsreader. It is a convention that reporters say who and where they are at the end of an item, and hand back to the studio. You can work out why these conventions are used, what effects they have on our views of the news programme as a whole.

1.7 Authoritativeness

This lies in the image presented by styles of news presentation. The popular press does not seek this upmarket image of being an authority on news about the world. But the quality papers do convey seriousness in their relatively print-heavy front pages and discrete headlines. It is television, however, which especially seeks to assume the mantle of authority through elements such as the dress of its newsreaders, its reporters on the spot, its up-to-the-minute information. This image is important because **it gives the news operation a kind of power – the power of being believed and trusted.**

1.8 Authenticity

News operations, especially those of broadcasting, like to enhance their trustworthiness and believability by appearing to present news 'as it really is'. The use of actuality footage, of reporters in real locations, of statistics through graphics, supports an idea that news we get is about 'the truth'. Pictures, whether in newspapers or on television, can be particularly influential in this respect – the cliché that if you see it it must be true. News editors will pay money to send news teams to cover an event or just the background to a story, even though they might be able to cover the story without location work, or could buy in material from an agency. Because 'we were there', because one can see the place where events took place or the people who were touched by events, then what they say about the story acquires credibility and authenticity. The 'Dunblane Massacre' story, where young children were murdered at school by a deranged man, is an example of this. The event was over. It served no real function of information to have pictures of the school or interviews with residents, except to bolster the credibility of the news organizations involved.

1.9 Experts

The use of 'experts' in news operations is also part of their image of authority. I use the word 'expert' in a qualified way because it is as much an idea that news people wish to promote as a straight fact. In other words, **they like to use and refer to experts in order to enhance their own credibility**. It is common to refer to reporters as being the 'consumer affairs reporter' or 'our correspondent in Beirut'. It is common to see experts in almost anything wheeled on to television news reports to express opinions. I am not saying that there is no expertise. But there is less than is suggested. The question is whether experts do substantially add to understanding of the story by being there in person or by being billed as experts. Most of what they are doing in at least some cases is to contribute to the credibility of the news operation.

1 .10 Consensus

This refers to broadcast news only. It defines a tendency in the treatment of social and political issues to deal with them **as if the middle view was always right** and was the agreed view. Clearly this cannot apply so much to newspapers because they are blatantly tied to the views of their owners and are in business to make money. But broadcast news is not set up to make money (though it can influence advertising revenue through the ratings it generates). It should not support political views or any partial views at all, because of the terms of the BBC Charter and of the Broadcasting Acts.

But broadcast news does support this consensus. In the case of stories about kinds of dispute it will always imply that a compromise is the fair solution and is good for everyone. This is not necessarily true.

1.11 Editorializing

This is the **inclusion of an editorial view or opinion on news material.** Newspapers have specific sections which express such opinions, which may support political parties' views. Broadcasting cannot do this, mainly on the assumption that it is somehow more influential as a medium and that by contrast newspapers can at least offer a choice of views. This idea of choice in the press is itself disputable. What is more to the point is that editorializing may happen covertly. This leads one into the area of bias (see below). For example, suppose there is on radio news a story about a possible takeover by one water company of another. It is covert editorializing if there immediately follows an item on water pollution perpetrated by one of the companies. A point of view is implied by the association of one item with the other. If the two items are right next to one another, like two pictures in a magazine, then this is an example of **juxtaposition.** Again, it may be that broadcast news deals with material about Iraq in terms critical of the regime. Britain was involved in a war against that country in the early nineties, so we are not surprised to receive news which is selectively critical. But the fact is that we can get an editorial point of view on the event and related issues. Suddenly neutrality is abandoned as an ideal.

1 .12 Impartiality

This relates to editorializing because it refers to **the idea that broadcast news doesn't take sides.** This is an ideal to be striven for in the way that stories are covered. Editors believe it to be a quality of their programmes. In general terms it is true that broadcast news is not partial to one political party or another (though party leaders have disputed this at various times). But what I have said above indicates that news is not totally impartial about everything. Yet it is enjoined to be, by various broadcasting acts and through internal advisory documents, 'each authority . . . must ensure that their programmes display, as far as possible, a proper balance and a wide range of subject matter, accuracy in news coverage, impartiality in matters of controversy' (1981).

1 .13 Bias

News may incline to one view rather than another, to one interpretation of events rather than another. The issue of news bias is always being debated (see Chapter 7). Newsmakers talk freely about their lack of bias. But everything I have said denies this – without saying that bias is extreme. **No communication can be totally neutral.** News people write out of their background and their beliefs – their ideology. Newspapers are biased by definition because no communication is neutral or value free. They frequently declare their support for a given view on political and social issues. Bias in broadcast news is less obvious but well documented in critiques of these operations. It has been pointed out that on a number of occasions broadcast news does implicitly bias its handling of trade union disputes by giving more time to management views than to those of the union, or by showing the union members as being excitable and disruptive as opposed to the calm talking heads of management. The degree and significance of this bias is something for you to follow up in your course work.

1.14 Selection and Construction

News, like any kind of media product, is the result of a process of selection and construction. Items are selected in or selected out. Newspapers or news programmes are an artifact that is put together. In effect, meaning is constructed into them. The meanings do not just happen to appear, they are there because someone made them. There are various aspects of news making which reveal how conscious is this making of the communication. For example, the reporter or newsreader interprets events for us. As soon as they talk about 'confrontation' they are actually interpreting what has happened. They are asserting that there has been confrontation, where someone else might have talked about 'disagreement'. The fact that we never see the camera crew on television helps construct a meaning which suggests neutrality and truth. We are not made aware that someone was there choosing the camera angles and indeed the subject matter. Sometimes this construction is very deliberate, as when photo opportunities are set up for celebrities. It is certainly argued that people will perform for the cameras, so that the news event is no longer the real event as it would have been. And obviously the whole process of editing written text or editing news film is a means of constructing a view of the original event. So the concept of construction draws attention to the fact that communication is created. News is created. It follows then that a student of the media must look at why this happens (to sell the programme or paper), how it happens (through an array of devices), and what effect this may have on the audience's view of the particular event or story and of the world in general.

1.15 News as Entertainment

When one attends to the fact that news is a construct, then one is more or less saying that **there is little difference between news and fiction, which is also**

about constructed stories. The very phrase 'news story' is revealing. It actually suggests that it is something made up. And stories are about entertainment. One can see that at least some news stories have entertainment value, when they are about crime and disaster, when they are about scandal and tragedy, when they centre on the human interest angle. People readily talk about human drama in news. But drama is at least associated with fiction.

What is more, there are dramatic devices in the unfolding of a news story. There are heroes and villains; the reader is denied information until later in the story. The whole programme in broadcast news is constructed with peaks and troughs in the relative excitement of the news items. Commercial television organizes this excitement around the advertising break, just like a soap opera might. Tabloid newspapers are in no doubt that news must be entertaining, to sell papers. The headlines, the selection of stories, the telling of stories, all contribute to this entertainment quotient.

1.16 News and Discourse

Once more the idea of discourse appears because discourses abound in media material. News may represent any number of discourses through its stories. But there is a discourse of news itself. There is a particular use of **visual and verbal language which produces special meanings about the idea of news itself.** The ideas of authority and authenticity already discussed can be said to be two meanings we get through the discourse of news. News (see also News Values) gives us a meaning that the activities of those who have power in our social structure are of more importance than what is happening with those who don't have power. News actually reinforces this power and definitions of fame through the working of its discourse. So it will for example prefer a body shot picture of a pop star arriving at an airport on a flight which has been delayed by terrorist activity to a general view of tired and anxious members of the public arriving on the same flight.

However, whether one is looking at news or other material, Stuart Hall draws attention to the fact that in the process of communication there is a distinction between meanings at the point of encoding and those at the point of decoding. If news has a discourse then the structure of meaning which is in the minds of those who put it together may not be the same as those in the minds of those who read, view, listen to the news. I have talked as if the way that the language of the discourse is used, and the meanings it produces, are some absolute 'out there'. But it isn't that simple. 'The lack of fit between the codes has a great deal to do with the structural differences of relationship and position between broadcasters and audiences' (Hall, 1993). So, to make things even more complicated, what I understand by the idea of news, what I make of a news programme, may not be quite the same as, say, a 'young audience' makes of it. This does not invalidate points made about features of the language of discourse – the significance of the face on half-body shot of the newsreader. But it does remind us that we have to be careful about generalizing about the meaning of texts.

1.17 News – Social Reproduction

Ideas about discourse and ideology are part of the notion that **news operations in particular help construct social reality.** The meanings in discourse, the values within ideology are part of this reality.

> The social reproduction thesis . . . is always based on the assertion that members of the audience obtain from journalism information which will tend to support an ideologically loaded view of the world; one which will contribute to the reproduction of an unequal and fundamentally antagonistic social system without dysfunctional conflict. – McNair (1994) in *News and Journalism in the UK*

But there is a question raised as to whether this is quite true. Can the construction of social reality be seen simply as a reproduction of the dominant ideology, where the news machine expresses ideas – about values, about social relations, about power, which we accept uncritically? Perhaps the news does frame off our view of the world to a fair extent, but perhaps it also to an extent raises consciousness of issues, is sometimes critical of dominant ideological positions. Perhaps it does allow us some room to make oppositional readings of its texts, if not much room. After all, some news material does take a critical stance on issues of wealth, class and privilege. There is something called 'investigative journalism' which at times takes on the Establishment.

On the one hand the notion of construction and reproduction can be seen in phenomena such as the 'moral panics' which news media create over events and issues such as those to do with crime. The call for further gun laws after the Dunblane Massacre was an example. It is ideologically 'conventional' to call for more control, more power. On the other hand there was no such panic over the court case which focused attention on the dismissal of gay personnel from the British armed forces. Certainly some views were expressed that made Attila the Hun look like a pink liberal. But other news coverage gave space to make a critical reading of the unsubstantiated arguments of the Ministry of Defence.

1.19 Technology and News

At the end of this special study of news and related ideas it is worth looking briefly at a few aspects of new technology (NT) which have made great changes in how news is gathered and presented. It contributes to news values and qualities, such as immediacy and actuality.

Technology contributes to news messages in a number of ways. The advent of **electronic news gathering** via video tape and satellite transmission back to the newsroom has enhanced the emphasis on up-to-date news. In the case of television this also means further emphasis on the value of visuals. The audience expects to see up-to-date pictures. The news makers make every effort to show recent pictures or footage. This was evident during the Gulf War when the evening news sometimes showed film from the aircraft of missile attacks made the same day. The use of electronic displays, graphics and captions have added to this visual emphasis and a sense of drama in the case of television. The globe has shrunk selectively as some parts of it are easily available to satellite links, emphasizing the effect of immediacy and actuality. However, to an extent NT has created a sharper line between what is available and what is not. China is one of the biggest countries on earth, but denies access through NT for the most part, and so is a place little seen.

Television news is now marked by the dramatic use of **electronic displays**, driven by the ubiquitous computer. News items drop in mobile graphics, captions and satellite links, almost without pause.

From this it is a short step to the **computer-controlled news studios** which are now in use. These can be run by one person or can even be run at a distance. Cameras are controlled remotely; news sources are tapped into. Such studios can either be slaves to the main news operations, or are in use by satellite television and international news providers.

The use of **electronic compositing** of material in newspapers has helped them update their material quickly because it is relatively easier than it was to change page layouts. In effect this is like using a more powerful version of the computer that was used to write this book. At the same time, technology has made colour photographs relatively easy to produce, perhaps dangerously blurring the line between newspapers and the magazine format. This happens because companies now use electronic process cameras to record images in the production process, and use computers to control **electronic imaging** of the recorded material.

2 ADVERTISING AND PERSUASION: COMMUNICATION AND INFLUENCE

This section deals with some definitions, some terms and with techniques of persuasion in particular. It explains ideas which in turn help describe how advertising may influence us, and how it shapes our culture and society.

Advertising is not a form of communication but a way of using forms of communication to achieve effects. Because modes of visual communication are used so frequently in advertising, this section also deals with image analysis, adding to what was said in Chapter 2.

You should also understand what this section will not do, because there is a great deal of other material around which it is pointless to duplicate within the confines of this book. It will not deal with the mechanics of the advertising industries and the production of advertisements. It will not make extended analysis of advertising material, nor of verbal techniques.

The reason that advertising is so frequently dealt with in Media Studies is that it is the one type of product which nakedly supports the commercial values of the media. The income it generates underpins all media. Without their income from advertising, popular newspapers would cost 50 to 60 pence each. Without advertising there would be no commercial television or radio. Without advertising most magazines would double or triple in price. And remember that there are many, many ways of advertising which we take for granted, but which would profoundly affect our environment if they disappeared – posters in the street, material on shop counters, material that comes through our doors. The economic effects of advertising are enormous. Whether we like it or not, advertising does create jobs by creating demand for products. Whatever the arguments about its effects, there is clear evidence that advertising does boost sales of products when it is present, and that product sales do drop after a while when it is absent. So we have to take very seriously such communication which affects all our lives in different ways.

It makes no bones about trying to persuade us, to have some effect on us. **It is very intentional communication.** Its purposes are clear. It is often very well constructed (and possibly effective) communication because the people who create advertisements have invested a great deal of time and money in finding out how its messages should be best treated to have an effect.

2.1 Some Basic Background Terms

Advertising is paid-for persuasive communication such as appears between programmes in broadcasting, or point-of-sale material in shops.

Publicity is free communication which may still persuade the audience to judge the product or service favourably. An example would be a film star taking about their latest movie on a chat show.

Marketing is the promotion of products (or indeed of things like banking services or a sale of shares in a company). This uses any device to promote its product, such as a staged event, as well as the media.

Campaign is co-ordinated advertising and publicity across the media and over a specific period of time. It is organized according to a specific schedule. A campaign uses a variety of means of communication at various times in order to get to a target audience and to reinforce its main message.

Target Audience is the particular audience chosen for the product or service, defined in terms of gender, occupation, disposable income and socio-economic grouping.

Sponsorship provides financial support to something like sport, or public service television in the USA, in return for which the sponsor gets their name spread around (for example on racing cars).

Product Placement uses the product within media material, often film or television drama, so that it is seen but not nakedly advertised. We have been used to seeing items such as Ford cars featured for years. But now placement of all sorts of items, including things like soft drinks, is big business and is paid for.

Brand Image is the impression which a particular brand of a product leaves in the mind of the audience. This impression can be dominantly about traits such as humour or classiness or value for money.

Copy is the writing created for advertisements.

Display Ads are the large advertisements in newspapers and magazines, usually with a picture and in some sort of a box.

	£
Daily Mail full page (black and white)	26,208
Daily Mail full page (colour)	37,800
The Daily Telegraph full page (black and white)	37,000
The Daily Telegraph full page (colour)	46,000
The Sunday Times full page (black and white)	48,000
Radio Times full page (black and white)	13,700
Radio Times full page (colour)	18,500
Just Seventeen full page (black and white)	4,900
Just Seventeen full page (colour)	7,930
Edinburgh Herald and Post full page (black and white)	3,080
Carlton 30 second weekday peak time spot (7.26 pm–11.30 pm)	12,000
Grampian TV 30 second weekday peak time spot (6.00 pm–11.00 pm)	1,250
30 second spot (each day, one week) in London cinemas (372 screens)	26,650
30 second spot (each day, one week) in Lancashire cinemas (162 screens)	8,587
BRMB (Birmingham Radio) 30 second spot, Wednesday–Friday (11.00 am–7.00 pm)	120
Virgin FM (London) 30 second spot, Monday–Friday (4.00 pm–8.00 pm)	150
Virgin Radio (AM/national) 30 second spot, Monday–Friday (4.00 pm–8.00 pm)	300

Source: Reproduced from *Student Briefing* No. 6, September 1996 by permission of the Advertising Association, UK.

Fig. 6.1 Examples of British advertising rates

The media fix their advertising rates according to the size of their audience and its age and social profiles. The rates are highly negotiable depending on numerous factors including possible large discounts. Here are some examples of mid 1996 rates.

Rate Card is the table of how much it costs to advertise in a given medium. Rates in television, for example, are extremely complicated, depending on factors like time of day, time of year, national or local screening, whether you are a local advertiser or not, as well as the numbers of your target audience who are likely to see the product.

2.2 Categorizing Advertising

One value of doing this is to remind yourself of **the range of means of communication available** and of **the range of things that can be advertised**. For example, people tend to think immediately of products and of high-profile media such as television. But posters are also a successful medium, and again people use unusual means such as balloons and matchboxes to put over the company name. And the line between publicity and promotion is blurred, if we use the idea of 'paid for' as a criterion. For example, launch parties for films or other products are quite common. They are clearly distinct from paying for a radio spot, but again someone has to foot the bill for the party. Another important category is the mail shot which comes directly into our homes as much as television. There are also the small ads in newspapers, which though they look very different from an expensive display ad selling new houses, are nevertheless paid for and are trying to sell a product of some sort.

2.3 The Purposes of Advertising

Similarly advertising can be categorized by its purpose. It isn't just about selling objects. The Government is often the top spender on advertising with a range of purposes, from warning the public about the dangers of drink driving to creating a favourable impression of the electricity industry prior to selling it into private ownership. When a company like BP spends hundreds of thousands of pounds on television advertising, it isn't trying to get you to buy petrol. It is trying to get us all to see BP as an impressive organization, doing good things like providing jobs and protecting the environment.

Even within a purely commercial environment, advertisers can have a range of purposes:

- to create awareness of the product or service to reassure existing customers about the quality of the product
- to reassure the trade and the sales force
- to grab a bigger share of the market
- to hold on to an existing share of the market against competition.

2.4 Advertising as Communication

Advertising provides us with a miniature of the basic process model of communication. It has a triple **source**, which is the original organization, the creative consultancy which creates the advertisement and the medium which actually projects the message. It is created because of commercial needs. It is **encoded** for specific media. It uses specific **media or channels** of communication. It contains messages of both an informational and a value-laden nature. The messages are treated so that they will be attractive to the audience. The **meaning** of the message(s) has to be decoded by the audience. The audience is carefully defined (targeted) as the receivers of the communication. There is **feedback,** notably through purchase of the product or service (or lack of this!), and through market research. All in all, advertising is a potent subject for Media/Communication Studies because it can help you understand ideas about meaning, about values, about ideology, about culture, and indeed any of the topics dealt with in this book. The next few sections look at some of these ideas.

2.5 Advertising and Representations

Advertisements like genres, are prone to use stereotypes (see Chapter 5, Section 3). **If they use stereotypes then they also tend to project the value messages contained in those stereotypes.** These representations of people are attractive to the makers of advertisements because they are instantly recognizable and so provide a short cut into the 'storyline'. These images of people are part of the attraction of the advertisement – and it needs to attract attention rapidly. They make the 'story' of TV advertisement or full-page magazine advertisement containable within a brief space (perhaps 20 seconds in the case of broadcasting). They carry the audience along with the main message of the advertisement because many of them agree with the value messages in the stereotype – largely because of the way they have been socialized. This is an unpalatable fact to those who object to the messages in the images. Part of the problem is that the nature of stereotypes and the construction of advertisements positively discourages us from stopping to think about them. Particularly if the advertisement invites us to have a laugh, it might seem ill-humoured to criticize it. However, a good student should stop to think about meanings and effects not least when the same stereotypes are reinforced by being repeated. We may remind ourselves that representation refers perhaps to ideas and institutions, to a version of reality, all of which may be associated with representation as it refers to people.

So you should ask yourself what it means when women are so plainly absent from advertisements relating to banking services. The bankers are almost always male. The customers, even the newer and younger target audience, are almost always male. This is a view of the world which does not see women as senior bank workers (statistically they are not!); it is a view of the world which does not see young females as potential earners. Such communication, supported by myriad other examples, is likely by accumulation to convince young females that they are never going to be economically powerful, that they are never going to make it in the banking profession. In this way stereotypes can contribute to a self-fulfilling prophecy, a vicious circle in which people end up behaving in stereotypical ways.

These stereotypes in television advertising are quite often set within a narrative borrowed from genre. See if you can spot the miniature soaps, quiz games, Westerns, war films and so on. Remember that in borrowing the genre the advertiser also borrows all the themes and messages associated with the genre. Holsten Pils is famous for borrowing movie clips. When it borrows an escape scene from a war film, however that is sent up, it is also borrowing something of the male macho associations of such films.

2.6 Advertising and Lifestyles

For some years now **advertising has sought to sell not just images and values but also the whole lifestyle within which the product or service exists.** These lifestyles are not invented but are, rather, sharpened versions of our own lifestyles and aspirations as obtained through market research.

> The function of advertising is to deliver audiences to the market . . . mass communication allows them (advertisers) to build up particular class, age and gender-related constituencies whose habits can be recognised . . . so when market research uncovers new social trends advertisers are feeding back to us versions of ourselves. – Hart (1991)

From another point of view it is all about giving us a story which we can drop into, a life which we can inhabit, and which of course is better than our own. It is no accident that in this same period there has emerged a kind of advertisement which tells a story, perhaps one which runs over a number of adverts. The story within which the Renault Clio exists is one of romantic adventure and Gallic style. It is a world for young women, and reflects the fact that females are now as important as males in terms of car purchases.

The lifestyle is one which gives pleasure and security, and which reinforces certain values, notably that consumption is OK and can even be squared with politically correct values such as a concern for the environment.

2.7 Advertising and Culture

Advertising in many ways represents key elements of our culture. It is central to our culture and to the ideology behind it because it is about consumption and materialism. It is one of our central beliefs (discuss it with your friends!) that it is a good thing to buy and own goods. These goods can make you happy, can enhance your status, it is believed. You are what your Porsche says you are, and this is good. Since one main business of advertising is to sell goods for consumption, **it is maintaining and promoting beliefs central to that culture.**

Some would say that it even helps to distort them and creates the contradictions to which I refer in Chapter 7. For instance, part of our ideology is about helping other people (perhaps nurtured by religious convictions). But there is at least some contradiction between the idea of being unselfish and helping others and apparently selfishly pursuing the acquisition of wealth.

Culture is a very complex idea which includes the artistic and creative sides of the society. So here are a couple more ideas for you about this powerful use of communication. One is that in a sense advertising is its own art form within our culture. This may seems odd, given the fact that advertisements shamelessly rip off Western art – showing a Rembrandt portrait smoking a cigar, for example. But the fact is that, just as people discuss other media products as a part of our culture, they also discuss advertisements – 'did you see the one about . . .'. Fellini's response to a reel of British TV commercials was to call them 'little masterpieces lasting one minute'. For example, there is a well-known cigarette poster advertisement which involves visual play on purple silk and the silk being torn by sharp instruments like scissors or knives. The symbolism is very peculiar, the meaning a good one for discussion. But still it is very much part of our culture, something that is not likely to mean anything at all in Japan. A related idea is that just as nursery rhymes are part of our culture – they refer to events and mythical characters that only we could understand – now advertising jingles are too. These are what children sing in the playground. They have become as much part of our culture as Bing Crosby singing *White Christmas* for the millionth time at Christmas.

2.8 Advertising and Ideology

I have said quite a lot elsewhere in this book about ideology, so it isn't necessary to repeat definitions and arguments. But it is worth pointing out that **notions of lifestyle and of culture tie in with ideology because they are all about peoples' values and attitudes**. Vestergaard and Schroder (1994) suggest that 'the most coherent, accessible version of the popular ideological universe can be found in the textual messages which people consume regularly, because they find pleasure in them.' People find pleasure in adverts, they are designed to be pleasurable. Adverts have for example shifted to take account of shifting attitudes towards gender and gender roles. We now hear relatively less about what it takes to be a good mum, and we do see something of young women in an independent role, even needing to use banking services. Having said this, there are still big underlying ideological imperatives that remain, for example the value of individualism. There is a kind of irony in the fact that cars, mass products on a vast scale, frequently pivot their advertising on appeal to individualism, being different.

It is car adverts among others that are referred to by Vestergaard and Schroder when they discuss a particular example of ideology – the ideology of 'the natural'. They talk about 'imposing Nature as a referent system'. They talk about putting the car in some beautiful countryside so that it accrues a kind of value from simply being there. Another common invocation is that of nature or naturalness as an 'ingredient in the product', in health foods for example. A third variation is where a product actually claims to be an 'improvement on nature' – hair tints for example. The important point is still that nature is a benchmark of value and approval. And finally there are advertisements which actually 'counteract natural processes' while still invoking naturalness. Here a classic example would again be hair tints that are meant to conceal naturally grey hair, but which claim to restore a natural look. Naturalness is clearly approved of in the ideological context, but its meaning and value is hijacked and grafted onto products which are often a long way from being natural in any respect.

2.9 Devices of Persuasion

Advertising is essentially persuasive, and so lets us see how communication can be shaped to persuade the audience. You should realize that the devices described below appear in all forms of communication in which persuasion is used. You probably try to persuade someone of something every day of your life. So you can apply the devices to conversation, group interaction, and so on. You should also realize that there is nothing inherently sinful about persuasion. When the word is used in conjunction with advertising, people tend to think loosely about brainwashing and sinister purposes. But there is nothing wrong with trying to persuade people not to drink and drive, for example. So if you are going to analyse advertisements for these devices of persuasion, don't imagine that you are uncovering some great plot. All you are doing is discovering how communication works. The question of what you are being persuaded of, what you might do or think at the end of it, is another matter. The implications of persuasion operated through one of the mass media are significant because of their qualities as media, because of what is vaguely called the power of the media. You can look back to the first two chapters to see what was said about these ideas.

Repetition: people tend to believe messages which are repeated. They take notice of them. Hence the habit of repeating brand names or catch-phrases in advertisements.

Reward: advertisements offer rewards in various ways. Free offers are naked rewards. But often the rewards are psychological rather than material. Advertisements for household cleansers offer you the reward of being a good housewife, or perhaps of being a good parent who cares for your children.

Punishment: the reverse of this may be that you are implicitly threatened with punishment if you don't buy – 'don't miss the opportunity of a lifetime!'. Or, if you don't buy this dog food you will feel really guilty because you are not doing the best for your pet.

Agreement with Values: this is a strong device, because the communication offers value messages which you are bound to agree with and then ties them to the product or service, so that you feel you also have to agree to buy this as well. You agree that it is sensible to plan ahead and provide for your old age, so of course you should buy this insurance.

Identification and Imitation: the device plays on the status of the source. In this case the advertisement uses, for example, a personality or type that you admire and respect. So you feel bound to order another pint of milk as the advertisement says, because X drinks it. Identification also plays on types of people. If you identify with and approve of the type represented then you may be persuaded to possess the product which they are shown to own and use.

Group Identity: people have a strong need to be part of a group. This can bring them security and status, as well as a sense of worth and their place in the world. So advertisements often persuade the consumer by showing an attractive group member and by offering membership of that group along with the product.

Emulation and Envy: in this case the advertisement offers you something like a way of life which you desire. You are shown a Carribean lifestyle (shown as enviable by previous media material), and so you feel that you should buy this pension scheme that will enable you to retire into the good life.

Needs: most of all advertisers are adept at appealing to basic needs which we have, and which motivate all our communication anyway. The needs for personal esteem and for social contact are very strong. Many advertisements for beer offer to satisfy these needs. You will become one of the lads and have a good laugh at the bar if you drink this beer.

Try looking for these devices in a range of advertisements across a range of media. See if you can pinpoint what it is about the image or the copy which actually triggers the device. And remember that advertisements may well use more than one device at a time.

You might also see if you can relate these devices to the classic AIDA model in which it is said that all advertisements have to grab Attention, arouse Interest, create Desire and lead to Action.

2.10 Visual Analysis

Advertisements often include visual elements: in newspapers, magazines, television, point-of-sale material, mail shots, and the like. By concentrating on the visual form I am not running down the importance of words. But education gives proportionately a lot of time to words, somewhat at the expense of visuals as communication. So this section may redress the balance a little.

In Chapter 2 there was a short section on image analysis as a method of studying media material. This focused on **image position signs** (where the camera is placed and places us), **image structure signs** (how elements such as lighting and focus affect our understanding), and **image content signs** (what the objects in the image may mean). These signs all work together so that the whole is more than the sum of its parts.

Visual analysis of advertising images will make it clear that they are selling values as well as products and services. It will discover the devices of persuasion referred to above. It exposes the use of stereotypes and the covert messages embedded in much advertising material.

There are one or two other concepts useful in visual analysis.

Denotation describes picture content. You should make a detailed description of every single item in a given advertising image before you jump to conclusions about what it means and how it works.

Connotation refers to the possible meanings of this content. Having got all the details, you should look at them separately and together in order to work out what the whole image really means, so far as you can tell.

Anchorage describes those elements which anchor the main meaning. Often it is a caption which ties down what the advertisement is about, especially when it does not have the product in it. A visual element may anchor meaning, especially if it dominates the picture in some way. Do realize that what is anchored may well not be the name or type of product. It may be some quality of the product, such as 'soft and gentle' for a brand of soap.

So visual analysis is very important when dealing with advertising as a use of communication. It reveals what meanings may be in the material and how they come to be there. Advertising is as significant within the media for its accepted use of persuasion, just as news is significant for its position as an informer. Both special cases in their own ways do a great deal to shape our views of the world. Within a limited space I have tried to deal with the SO WHAT question with regard to advertising. So it plays a vital role in the media. So understanding of some descriptive terms helps sort out what advertising is and what kinds of advertising there are. So it tells us a lot about how we use communication in general in order to persuade. So understanding of sign and process can be used to get at the meanings within and the significance of this special media product.

REVIEW

You should have learned the following things about news, advertising and persuasion from this chapter.

1 NEWS ENCODING AND DECODING

1.1 News is gathered and sorted from a variety of sources and is paid for like any other commodity.

1.2 News organizations decide on news topics for a day or a week through a process of selection. This is called agenda setting.

1.3 News is selected and handled according to values held by the news organizations. These values relate to content and treatment in particular.

1.4 News stories are handled from particular points of view with a particular theme in mind. This is called the 'angle'.

1.5 News has its own secondary code. The signifiers affect how we understand news items, and how we see news material as being truthful and authentic, for example.

1.6 The conventions of the code affect how news material is structured and presented to the audience.

1.7 News presented is seen as having authority, because of the signifiers in its code.

1.8 News nurtures an impression of authenticity, of truthfulness and actuality.

1.9 News material is mediated and given authority through the use of 'experts'.

1.10 Broadcast news tends to present social and political issues in terms of a consensus, or belief that the middle view is always right.
1.11 The presentation of opinions overtly in newspaper editorials or covertly in the way that news material is handled is called editorializing.
1.12 The various news machines believe in impartiality though they do not always achieve this in practice.
1.13 There are continuing debates about the issue of bias in news, especially in broadcasting. There will always be a degree of a bias because communication cannot, of itself, be neutral.
1.14 News is a process of selection and construction of meaning about its material, through various devices.
1.15 News is in many ways just entertainment, especially on television.
1.16 News has its own discourse, or special language and sets of meanings about what news is supposed to be.
1.17 News helps construct our social reality. It helps reproduce 'things as they are'.
1.18 News making and presentation has been much influenced by new technology, which tends to emphasize the value of recency.

2 ADVERTISING AND PERSUASION

- advertising is not a form of communication, but a kind of use of various forms of communication.
- advertising is crucial in Media Studies because it pays for so much material.
- advertising is intentional, persuasive communication.

2.1 There are a range of basic background terms which are useful for defining the topic – publicity, marketing, campaign, target audience, sponsorship, product placement, brand image, copy, display ad, rate card.
2.2 Advertising is not all about product. Other descriptive categories include service, company image, public information.
2.3 Advertising has a variety of purposes beyond trying to sell objects.
2.4 Advertising may be analysed effectively in terms of a process model of communication.
2.5 Advertising tends to use stereotypes to help carry messages about its product, service, etc. These stereotypes also carry covert value messages.
2.6 Advertising constructs whole lifestyles for its audiences, and sells these lifestyles.
2.7 Advertising material is part of our culture and represents ideals about that culture.
2.8 Advertising reinforces the dominant ideology in its value messages.
2.9 Advertising uses a number of devices of persuasion, main examples of which are repetition, reward, punishment, endorsement of values, identification and imitation, group identity, emulation and envy, arousal and identification of needs.
2.10 Advertising relies heavily on imagery. Image analysis is a very appropriate way of getting at the meanings in advertising material.

Activity (6)

This activity looks at the crossover between news and marketing.

The various tasks are in themselves concerned with writing and designing, and so develop some practical skills. In respect of treatment they ask you to be aware of how this inflects the message or meaning of the piece of communication. But behind all this is the sense that media products are not as distinct and separate as the media producers would wish us to believe. Categories are a convenience for the producers, a way of positioning the audience. Of course, audiences quite like to know where they (think) they stand. But critical understanding of product leads one to appreciate the relative falsity of distinctions.

News as Promotion

- WRITE A NEWSPAPER ARTICLE which would benefit the opening of a film of your choice – i.e. a story about a London première.
- SCRIPT A TV NEWS ITEM which would benefit the image of a pop star/group of your choice – i.e. arriving at or leaving an airport at the beginning of a tour.
- SCRIPT A RADIO NEWS ITEM about the anniversary of one of the new national radio stations – i.e. which is hung on a celebrity party of some kind, which is all about the success of this station.

Advertising News

- DESIGN A BILLBOARD POSTER for a national daily newspaper of your choice – i.e. as part of a national campaign to hold on to readership.
- STORYBOARD A TRAILER for Sky News – i.e. a progamme trailer on Sky itself to make Sky News attractive compared with the competition.
- STORYBOARD A TV ADVERTISEMENT for a magazine of your choice – i.e. one which is helping the relaunch of a national magazine for women.

Points to reflect on from these activities include the question of why news should need to sell itself anyway. News is news, or is it? In any case, why should news promote itself when, at least in broadcasting, it is a compulsory part of the programme mix?

And what is news up to providing free publicity for other parts of the media industries? Would it be possible and reasonable to ban newspapers and news programmes from providing coverage which has nothing to do with raw information about the media, such as takeover bids by media empires?

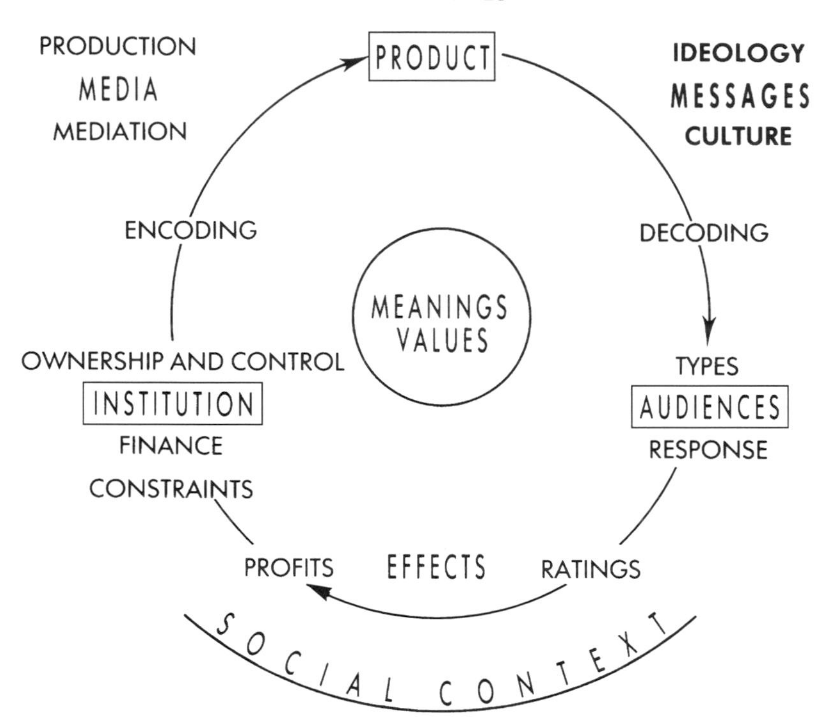
REPRESENTATION
INFORMATION AND PERSUASION
GENRES
REALISM
NARRATIVES
PRODUCTION
MEDIA
MEDIATION
PRODUCT
IDEOLOGY
MESSAGES
CULTURE
ENCODING
DECODING
MEANINGS
VALUES
OWNERSHIP AND CONTROL
INSTITUTION
FINANCE
CONSTRAINTS
TYPES
AUDIENCES
RESPONSE
PROFITS
EFFECTS
RATINGS
SOCIAL CONTEXT

7

Meanings

Did You Get the Message?

The answer is that you probably did, even if you didn't realize it. All of us are getting media messages every day. This book has already said quite a lot about what these messages are. But in this chapter I want to look more carefully at the kinds of message that are coming through, and at ways of getting to the more covert messages in media material. We need to get beyond seeing the obvious information in a programme or paper as being the message, and understand that within it **there are implied beliefs and values which are also messages**. A newspaper may give us a front-page message about a helicopter crash which kills 20 people in our own country. But what it also does is to give us covert or implied messages about this crash being more important than the small inside-page report on the collapse of a dam in China which kills 50 people. To get into all this we have to start off by exploring some ideas about messages and meanings.

1 MEDIA MESSAGES – THEIR MEANINGS

1.1 The Meaning is the Message

In any example of communication, **the message is what is meant by what is said or written or pictured.** This may seem fairly obvious. But it is surprising how people confuse the two words, meaning and message. The distinction is usually once more in terms of what is obvious and what is hidden. For example, a certain advertisement for a bank is seen as carrying the message – 'if you are young open a bank account with us and we will see that you are all right.' But in fact the meaning of the whole advertisement can be a great deal more complicated. One part of it is that bank managers are of course male. Another is that it is quite easy to borrow money from a bank. There may be more. But at least you should now understand that there is more to the meaning than may first appear.

1.2 Multiple Meanings

There are usually multiple messages in any example of media material. One picture can tell us a great deal; so can one magazine, one programme, one record. There is always more than one point to a story. There is always more than one point of view about some political event.

1.3 Lifestyles

One comes back to the idea that **many messages in the media are about our lifestyle, about the groups we belong to, about our self-presentation, about our beliefs and values.** Think about magazines for young females. They certainly have messages about a lifestyle which is dominantly to do with dressing up, having a good time and being interested in young males. They are about the idea that the young female readers are somehow separate as a group from other females in the population. They often (but not always) assume that young females go out as a group – until they meet a certain young man. They carry a lot of messages about self-presentation, about how to look, since they carry many articles and images which refer to hair, clothes, make-up, and generally to the idea of being looked at. But then it follows that they also carry the belief messages that young women ought to pay attention to their appearance, that they should expect to be looked at. It also follows then that they carry value messages about the importance of personal image. Put another way, they are saying that young women are valued according to this image.

1.4 Naturalization

One major message that we receive through the media is that media messages are 'naturally' correct. The media are so seductive in the accomplishment of their communication, especially entertainment, that they gloss over the covert messages. They deliver an assumption that this is the way things are, and should be, these are the ways that things should be done, that what the media say is basically satisfactory and correct. Broadcast news and newspapers carry a view of the world that is never questioned, that is also made natural. It is a view which naturally shows politics in terms of conflict, which refers to government and opposition, which naturally assumes that people have to be on one side or the other of some political argument rather than perhaps agreeing with parts of three or four views. This 'naturalization' works against uncovering what the media may really be telling us, against questioning what they do tell us.

1.5 Intentionality

It is important to realize that **whatever meanings we may find in the media, they may or may not be there by intention.** We have to steer a course between paranoia about what 'they' are trying to do to us, and a naïve attitude which assumes that all messages appear by chance, or that everything is written and produced for essentially benign reasons. It is absurd to suggest that the media producers are intentionally trying to feed us a diet of stereotypes through drama and comedy. It is equally absurd to assume that they are not aware of these stereotypes. And it is stupid to ignore the fact that we do receive value messages through such stereotypes. It is necessary to **understand that even unintentional messages are still the responsibility of those who construct them** into material (however ignorant they may be of what they are doing). And it is important to realize that some messages really are there by intention. Most messages in advertisements are there by intention. One would have to be very stupid to spend thousands of pounds on cat food commercials and not realize that one has intentionally encoded messages about cats being females' pets – not least because females do most of the shopping, for pet food among other items.

1.6 Preferred Readings

The way in which the message is handled in a given medium can cause us to prefer one reading of it, one meaning, above other meanings. In effect this is a kind of bias in the way that the message is put across. Once more, this preferred reading could come from a conscious initial intention to make the audience see things in a certain way. Or it could come from an unconscious but firm way of seeing things, which the makers of the communication put into the material without realizing it.

Conscious and persuasive mass communication is an obvious example of the intentional kind of preferred reading. For example, a holiday firm might mail-shot households giving them colourful attractive leaflets about beach holidays. On the surface the leaflet just tells us about places and prices. But mainly it tells us about preferred places, it tells us to prefer that company above others, to prefer beach holidays above others. Obviously this is done through a combination of written information and selectively shot pictures, to put it briefly. There is no possibility of understanding holidays and beaches in a neutral way.

But there are also preferred readings to be seen in material which is apparently more factual and objective. To take a reasonably provocative example, we have seen in recent years documentaries about the destruction of rain forests and about the effects of this. If you can bring to mind such aerial shots of destroyed swathes of forest, of miserable-looking dispossessed Indians, or the voice-overs which comment on the possible effects on wildlife, on the climate of the Earth, and so on, then what you are remembering are the cues which prefer you to read the programme and the images as a negative view of clearing the forests. The programme makers prefer you to read their 'story' this way. They leave out any views of the benefit to the economy of the country concerned. There is no question that we are being shown pictures of destroyed forest just out of general interest.

This notion of a preferred interpretation of the material extends especially into news, supposedly the most neutral aspect of media production, the words and pictures which support headlines like 'Travel Chaos' or 'Misery for Commuters' when there is a some kind of transport industrial action clearly prefer us to see that action in a negative light. The news media could prefer the view that this is a healthy exercise of employees' power in support of their rights; or the view that it is all a temporary blip on the economic landscape, or that it is all a necessary consequence of firm management policies. But, in fact they prefer us to read the material in another way.

1.7 Accumulation and Repetition

Media messages achieve their significance and their effect, it is theorized, by sheer accumulation and repetition of material. A message such as 'A Mars a Day helps you Work, Rest and Play' **accumulates its impact over a period of time** (in this case nearly 60 years), and across a variety of media. Similarly, a message like, 'most doctors are males (and should be)' does not come just from one situation comedy, perhaps, which shows only male doctors, but also from an accumulation of various media sources which say the same thing – romantic stories about the handsome young doctor; advertisements for nursing staff which show a male doctor; radio interviews with representatives of the doctors' professional association, who are usually male; etc.

1.8 Messages, Meanings and Ideology

Issues

One category has to do with **issues of the day, which the media themselves often define anyway**. One current issue is whether or not nuclear power is desirable. You could follow this up through the press only and acquire a useful dossier of messages, overt and covert. Overt messages include the various advertisements from special interest groups such as Greenpeace and the Nuclear Energy Authority. These take one side of the argument or another. But you could also look at various articles and opinion columns in newspapers. Sometimes the messages and positions on the issue are interestingly oblique. For instance, there have been articles about the incidence of leukaemia around power stations. Does the relative frequency and scale of reports about this problem in the quality press suggest a basically anti-nuclear power message? Does the relative absence of such comment in the popular press suggest lack of interest, or a pro-view on the issue?

Normality and Deviancy

Another set of messages **in effect define for us what is 'normal' in our society and what is inappropriate or wrong.** Writers like Stuart Hall in *The Manufacture of News* (Cohen and Young, 1989: Constable), Stuart Chibnal in *Law and Order News* (1977: Tavistock), expand on this idea considerably. But, to put it simply, the media are a significant influence in defining for us what we should regard as normal behaviour and normal people, and what is abnormal behaviour by people who in effect are labelled deviants. This is easy to see in terms of gross criminality. Child murderers are universally depicted as deviants, and murder is abnormal. Again, it isn't just news or factual material which conveys this. We get these messages from a range of material, including drama.

But you should realize that sometimes the messages which help define our view of the world may be more questionable in their validity than is immediately obvious. For example, in Britain, there are Welsh nationalists who burn down houses owned by English people. The media clearly tell us that these Welsh people are **deviants**, that their behaviour is not normal. But there are at least some Welsh people who do not agree with this. They would say that such people are patriotic, not criminal. Another example has to do with lifestyle. Those members of our society described loosely as Travellers are also generally depicted as abnormal and deviant in respect of their lifestyle and beliefs. They are often dealt with critically when they settle on other people's land or because they may not give their children a regular education. But from another point of view their relative lack of interest in possessions and a material lifestyle could be said to have a positive dimension – one that isn't often promoted by the media. People who prefer to sunbathe in the nude have been represented as mildly deviant or 'odd' in our society. People who support environmental policies above others have been similarly represented as eccentric and untypical. It is interesting that their brand of deviancy has been revalued by media in many parts of the world as it became obvious that Green parties were entering the mainstream political process and were numerically significant.

■ *Culture and Subculture*

The media help define what our culture is, and what subcultures are in relation to mainstream culture. They define views of and attitudes towards these subcultures. To some extent, minority programming in broadcasting does this, rightly or wrongly. For example, radio programmes such as *Does He Take Sugar?* for the disabled, define those groups as subcultures within our society. In a general sense all targeted magazines help define subcultural groups – for instance computer games magazines aimed at young males. To this extent, the media identify and maintain subcultural categories and divisions. They may deal with specific issues such as the notorious debate about Salman Rushdie's book called *The Satanic Verses* to which Muslims deeply object and over which they have taken to the streets in protest. In this case, there have been innumerable messages in all the media taking different views about British Muslims' right to this protest, and about the degree of their separateness from what is called mainstream British culture.

■ *Gender, Age, etc.*

In effect these have already been dealt with under the heading of stereotypes. But it is worth reminding ourselves that there is a **whole category of messages that has to do with putting people into 'boxes'**. The box of age is one in which there are dominant messages about eccentricity, physical and mental incompetence, conservative attitudes, dependence on relatives, a lack of understanding of younger people.

■ *Ideology*

One set of messages that we receive through the media is about power. These tell us, more or less directly, **who has power and who ought to have power.** If we look at television news programmes and the front pages of newspapers then

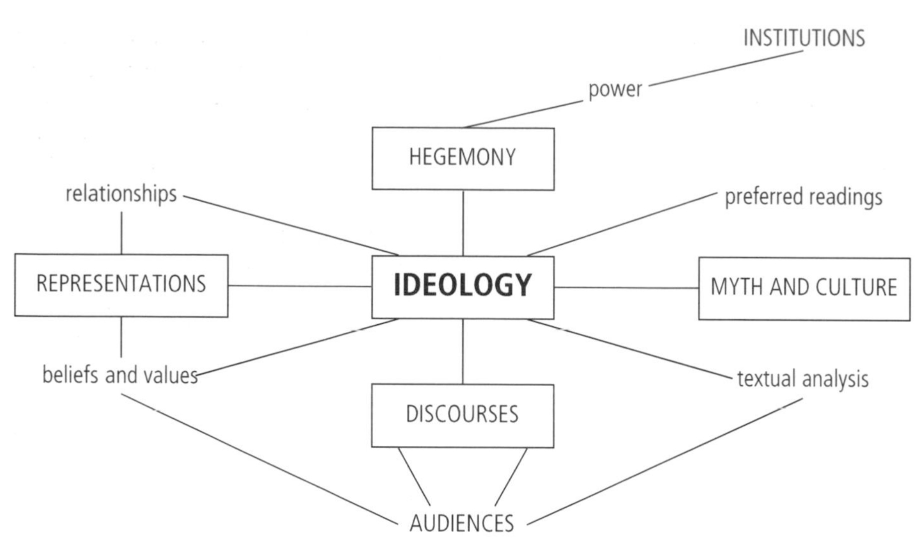

Fig. 7.1 Key concepts – Ideology

it is reasonably clear who has power. For the mythical Martian coming to Earth it would seem that more men than women have power; that these men are older rather than younger; that many of them are politicians; that quite a number are entertainers; that most of them are white; that a number of them are in charge of countries across the whole globe. It also becomes evident by contrast who does not have power: young people, most old people, poor people, female people, disabled people, black people. You can make your own analysis to see if you can add any points to this.

In this case, you might also reach such conclusions as: power seems to be about control of government, control of money, control of armed forces, and control of audiences.

The power we are talking about is exerted in relationships between people and groups. The politicians exert power over groups of people. The entertainer exerts power over the audience. Leaders of industry exert power over groups of workers. It is also worth realizing that we may see the media giving messages about power in more personal relationships. Or they may show power as being more personalized – between individuals. For example, films such as *Wall Street* or soaps such as *Roseanne* show power struggles in relationships. Sometimes this is about business – white males trying to win takeover battles. But sometimes it is about love and marriage – power struggles for affection, even possession of another person. Sometimes in such dramas we see men trying to own the women in the same way that they own companies – perhaps by 'bribing' them with gifts and a good time out. We may also see the women exerting power over men through using or withholding their affections. In this case, we are getting more messages about power in personal relationships, and about how it may be operated.

The important thing to realize is that such **messages about who has what kind of power over whom, are not simply reflections of the world as it is. By being given to us, they actually help make up our view of that world.** This is not to say that the messages are totally untrue. Nor is it to jump into judgements about whether the situation as described to us is a good or a bad thing. But it is to say that the very act of communicating is one in which a message and its treatment is selected, and therefore cannot by definition be neutral. It is to remind you that the very repetition of such messages in various guises, reinforces their meaning. You need to be aware of this, because in this way the messages themselves have power over you. You can only take back the power to sort out meanings and conclusions for yourself if you realize what is happening.

Discussion of power in the meanings generated by the media leads one directly to the notion of **ideology**. Ideology is a difficult word to deal with because there have been so many interpretations of it. It is also difficult because it has strong associations for some people – often the idea that it is all about left-wing politics. This is not true. Ideology is, somewhat simply, **a set of beliefs and values which add up to a particular view of the world and of power relationships between people and groups.** All sets of beliefs which have labels or titles are ideologies: Buddhism, communism, capitalism, Catholicism. These 'Isms' are often tied to particular cultures, but more than one may exist within a culture. There are at least some Russian people who follow the Russian Orthodox church and its beliefs, but who are also communists, just as some English people may be Christians and socialists.

All of us have some sort of ideology or view of the world and how it is and how it should be. Our ideology is formed by the culture in which we grow up. In particular, we are influenced by communication which comes from family, friends and school, and also the media. We are influenced by those messages which are about what we should believe in and what we should value as important. This is why any study of communication is likely to come back to value messages, because these become part of the way we think. They affect how we put together communication to other people and how we decode communication from others. Much of this book is about the process of communication through which the media project these values, and so form them or reinforce them (sometimes even question them). The final chapter in particular is about media effects, and so also about how such value messages may or may not be perceived by us, and how they may be taken in and used by us.

The dominant ideology is the dominant view of the world in a given culture. This is the one which the media offer us, for the most part. It will also probably be the view of the world that we hold. This overlap is not surprising, simply because the media seek to please the audience to get readers and viewers in order to sell their product. It is a simple fact, well known to advertisers, that if the messages within communication include those about values which the audience holds, then the audience is most inclined to accept that communication and to like it. The 'catch' is, as we have just said, that those same media help form those values in the first place. By the same token, they are not inclined, on the whole, to raise questions about those value messages, about the nature of the ideology, because this would spoil their success as communicators (in their terms). In my terms I suggest to you the reader that you are actually a better communicator if you are able to question how and why communication takes place, as well as what messages are being communicated.

The value messages about this ideology that we have in our heads are enormous in number. The most important categories are those that have to do with power, and this we have already said something about. If you were to write out a list of such messages as statements, it would be very long. I will only give a few examples of what I mean. You can find others by discussing the moral of various media stories. For instance, one message is that the family is a good thing, or that law and order is a good thing, or that it is OK to use force so long as you are on the side of the Law, or that it is good to save money and to own things, or that love and courage are good qualities – and so on.

■ *Hegemony*

This term relates to ideology because **it refers to the invisible exercise of power by those who run things over those who do not**.

One has to beware of falling into unfounded conspiracy theories. However, it is objectively unarguable that **relatively few people in most societies control (legally, politically, financially) the dominant institutions of government, the military, commerce, the media**. In Britain we would be looking at something like 20,000 people in Parliament and top jobs, who have a tremendous influence over the lives of the other 54 million souls.

Hegemony suggests that the influence does not have to be direct and obvious to be important. It suggests that it has something to do with class and gender (look at the gender and backgrounds of those in top jobs). It constructs the web of values that holds together our ideology.

Hegemony relates to the concept of **naturalization**, which describes the fact that the way media material is put together and the meanings which that material contains, are made natural. That is to say, it is assumed for example that women should be defined in terms of beauty, or that we should be concerned about the exchange rate. It is as if the media are saying, well of course this is important, of course this is true. Not only could one question definitions of importance and of truth, but one could also ask. Who says this is important? Or even who benefits from saying this is important?

In fact what the process of hegemony does is to naturalize ideology itself, and all the values and beliefs built into ideology. Hegemony is at its most powerful and most dangerous when people assert that things are naturally true, or even part of human nature. It has been a received truth that men are naturally better drivers than women – but insurance company statistics contradict this. Or that the process of law ensures justice for all – when clearly it brings more justice more easily for those with more money.

So from this point of view the messages and meanings within media products like newspapers and films are about some beliefs rather than others.

And these beliefs that are presented work in the interests of media owners, of advertisers, of something we call the establishment – rather than in the interests of ourselves, the audience.

■ *Discourse*

This concept is again linked with ideology and with the value messages that we get when we read magazines and watch television. The term describes **particular ways of using verbal and visual languages to produce particular meanings about the subject of the discourse.** In principle you could have a discourses about nearly anything. In practice one sees discourse as being to do with 'big' topics such as gender, or law and order. These discourses define what it is to be male or female; they define what law and order means. They are themselves defined by particular words, phrases, camera shots even. Language is not only used by the media of course – but it can be used with particular effect because of the nature of the media. So for example, if words to do with softness, scent and colour are used specially to describe females, then it is the media (e.g. magazines) which especially reinforces the meanings in the discourse. If camera shots and their point of view on females as subject matter particularly directs their 'gaze' at aspects of the physical form of women, then this repeated use of language also reinforces the idea that being a women means being a body.

In talking about discourses, Turner (1992) talks about 'ways of thinking that can be tracked in individual texts or groups of texts' – ways of thinking about gender. He also talks about 'wider historical and social structures or relations'. In other words, if one wants to understand current ways of thinking about gender, then one needs to take account of where these have come from, of how they may relate to thinking about gender in the past.

Discourses operate within ideology and they are part of it. The value messages of the ideology can also be the meanings of a discourse. The language of the discourse also communicates the value messages. There is a discourse about royalty and the monarchy in our culture. There are dominant ideological values relating to royalty – approval, patriotism, authority. Analysis of the special language used about royalty would lead one to exactly the same words.

Whether we realize it or not we use discourses to make sense both of our real world and of the world of the media. Different ideologies will have their own discourses which both represent how the relevant culture 'sees' the world and frames off how members of that culture make sense of their world.

1.9 Conflicts and Contradictions

I have already warned you against seeing media communication as some kind of grand conspiracy. I have also pointed out that if there are general truths about the content and method of communication, then these truths are never absolutely consistent. There are some alternative views of issues. There are unstereotypical materials. There are examples which question what the media themselves are doing and how. One of the interesting things you should come across in your investigations are messages which contradict or conflict with

Carol took over, guiding her to the squeaky sofa. 'Of course you know, Jenny, he's right in one way,' she said. 'The increase in crime, especially burglaries in areas like this, well it's appalling. I blame all those unemployed boys with nothing better to do.' She leaned forward eagerly, and then back again, frowning as she caught Paul's gaze homing in on her cleavage. 'Have you been in that estate up across the main road? All day long, absolutely heaving with people, all milling about with nothing to occupy them. We're heading for a revolution, that's what I think. They'll come pouring across that road and ransack nice, law-abiding places like this, just like Russia.'

God, she's mad, thought Jenny. Stark staring. Carol would never go across the main road into the estate, in case she was robbed both of her values and her valuables. How could she possibly have a clue what went on over there? But Carol, her spun-sugar hair jerking briskly like a rooster's crest, was just warming up.

'And the children, not a father present between them! No role models. I thank God we were able to send our boys safely off to a decent prep school where they can learn what's what. And there's a lot to be said you know, for joining the cadet force. Channels their energies.'

Into learning how to kill each other, crossed Jenny's mind, but she said instead, 'They could learn what's what here at home, surely, with both parents present?' suddenly keen to defend her own version of family life, 'if all you think they need is a father-figure?' Carol gave her a sly look, something conspiratorial, as if what Paul could teach her sons wasn't likely to be worth knowing.

'You know,' Jenny told her, 'the vast majority of crime by young people is done by boys. The same lone parents are also bringing up girls too. So surely it's the messages that are getting across to boys in society that are damaging, not necessarily their family structure.'

Carol, for a moment looked stumped, but soon rallied. 'Oh well, the girls, we all know what they're doing don't we? Getting themselves pregnant and straight up the housing list, that's what.'

Fig. 7.2 From *Pleasant Vices* by Judy Astley, Black Swan, 1995

This passage shows how discourses inhabit media product everywhere. They intermingle and work together, dominating the way that we make sense of the world. There is the line about the woman's cleavage, in which sexuality appears briefly, naturalizing an attitude about how men look at women. The conversation incorporates discourses about class and masculinity – for example a phrase like 'a decent prep school' belongs to a certain way of looking at the world. Interestingly, Jenny also argues with Carol, and produces an alternative discourse about the young and parenting. But then Carol's dominant discourses roll on again as she talks ignorantly about working-class girls getting themselves pregnant.

one another. These tell us something about the contradictions in society itself, and about how difficult it is to make absolute generalizations about the media.

For example, in a previous paragraph I referred briefly to ways in which we get messages about older people which are generally unflattering. There is an interesting contradiction here if you consider the fact that the media also tell us about old people, especially males, who are powerful and respected. Japan is run by men who are what we might otherwise describe as old-age pensioners. The USA had a popular President (Reagan) who held office well into his seventies. Newspapers abound with stories of politicians, financiers and film stars who are all old, but who are certainly not incompetent, as others of their age appear to be in dramas and comedies.

Trade unions which oppose the power of the state come in for a lot of criticism from government and from newspapers sympathetic to that government. Even demonstrating British ambulance crews and their unions have been criticized by government ministers. But government and newspapers have in the past praised a foreign trade union – Solidarity in Poland – which also instigated strikes against the State and brought down the government, because they did not like the communist ideology of the Polish government. This is not to condone other actions of the former Polish government, but to point out an inconsistency of views about trade unions.

Such contradictions of message are interesting in that they prove that **the media do not present a cohesive view of the world**, and useful because they help us be aware that **there are different levels of message in the media**. Messages about values and ideology are often at the deepest level. But they can have the most importance for us personally in terms of shaping our beliefs and in terms of affecting how we behave as members of our society. But before we deal with that important SO WHAT question: So there are these kinds of message – so what? So I can dig out messages – so what? – there is one more important point to be made about messages and the media. A lot has been said about messages from the media. Let us look at how we may or may not get messages back to the media organizations.

1.10 Access and Feedback: Did They Get the Message?

With all this talk of messages from the media organizations we must not lose sight of the two-way nature of the communication process. **The audience does feed back messages to the producers, though in a very limited way.** The messages are mainly approval or disapproval of the product. They come through letters to editors, phone-in programmes and general complaints to organizations. They may also come crudely through purchase or non-purchase of the product, through switching on or switching off. However, these feedback messages only operate within conditions controlled by the producers. So phone-ins are usually on a delay circuit which allows for censorship. Letters to magazines are often edited down. Most significantly, none of these channels of communication offers any assurance that messages will get to the most appropriate person, nor that the audience can in any way influence choice of material, its timing, content, treatment.

This lack of effective feedback in itself tends to influence the messages that we receive. The producers of media communication are answerable to owners and editors in the first place, not to the readers and viewers who ultimately pay for the communication. It is true, as we have already said, that the audience must be pleased by media material, otherwise they won't buy it. But this is a separate matter from the fact that the producers are not really accountable to the audience for the material that they make. Nor do the audience have access to the media. This is not to suggest that all members of the public have the time and the technical and presentational skills to make newspapers and television, for example. But it is not fanciful to suggest that some people and some organizations could work with media professionals to produce material that the public creates rather than just influencing what gets made. There are already examples of community radio and open-door type programmes on television. Dutch broadcasting has legal and financial arrangements by which organizations of various types – religious, political, unions – have rights to air time proportional to their membership. It is possible to operate media to which the audience has some right of access and therefore some direct hand in the messages which they receive.

2 MEDIA ISSUES AND DEBATES

We are going to look at a few of the major issues raised by Media Studies. In fact the media themselves occasionally deal with such issues. In so doing they also send the message that these issues are on the public agenda. This makes the general point that in many cases what we call an issue is only one because the media say it is. This underlines the fact that the media wield considerable power over our view of the world, of what is important or unimportant. What follow then are outlines of debates which you should be able to discuss, offering not only general opinion but also informed judgement from your course and from reading the earlier parts of this book.

2.1 Bias

The issue of media bias usually focuses on their news operations. Journalists usually conclude that they are neutral, especially broadcasters. One might beg to differ. For a start, since communication always comes out of the culture and values of the communicator, it is impossible for it to be neutral. Whether there is relative bias or not is another matter.

Bias in print news may be fairly obvious given the editorial positions of newspapers, the targeting of audience and the sometimes naked political affiliations of newspapers like *The Express*, whose Tory leanings are manifest in the material. Bias in broadcast news is more of an issue because this earnestly aspires to impartiality and neutrality. The provision of information is not in doubt. The inflection and framing of this information is something else.

Again, it has already been pointed out that selection out and selection in through the editorial process makes a kind of bias inevitable. News items are also biased through their priority in the running order, the time given to the item, its placing next to other items, let alone through the newsreader's script, which can make crucial choices (and inflections) when choosing between words like 'demonstration' and 'riot'. News programmes try to get rid of bias through devices like having both a Tory and a Labour politician interviewed on a given topic. But the very assumption of a two-party system or two views is itself a kind of bias. In any case the interviewer holds the ring and creates another kind of bias saying things like 'so could you sum up your position in three or four sentences'. Perhaps it isn't possible to do this. Simplification can be a kind of bias in itself. The valuing of pictures when choosing and handling a story can be another kind of bias. Clearly there is a fundamental issue here surrounding one of our most important sources of factual information about the world. Did the Soviets retreat from Afghanistan or did they make a strategic withdrawal? That depends on whose news you watch. Perhaps they are both biased!

2.2 Control

Who should control the media and how? This issue concerns the potential effects of the media (otherwise it wouldn't matter who controlled them). The debate ranges across arguments about who owns the media under what conditions, and about how the power of ownership should be moderated.

This power can be generally moderated by the Monopolies and Mergers Commission, which may be asked by Government ministers to rule on whether or not it is in the public interest for one company to take over another if that means the new company would have a big slice of a given industry or market.

So far this has had little effect on the ability of owners to control a number of newspapers or a number of radio stations. When it comes to **cross-media ownership** there are clauses built into the latest broadcasting acts. At the moment a newspaper group cannot have more then a 20 per cent stake in a broadcaster. However, relaxation of this rule is now being considered. And the Government has already relaxed the rules relating to ownership of more than one television company. So Yorkshire was able to take over Tyne Tees in 1993, and there are now rumours that Yorkshire itself could be the object of a takeover bid. In terms of commercial radio and television there is no argument that a degree of control is in the hands of the Independent Television Commission, set up in 1992 to replace the IBA. Its remit extends across to cable and satellite. It can for example give or take away the broadcasting contracts which allow companies such as Carlton to operate.

But there is also the matter of the nature of that control. The dominant model of media insitution in this country is a commercial one, with devices for checking and curbing misuse of power (through organizations such as the ITC). But should we look for alternative models of ownership or for regulatory bodies? *The Guardian* and *The Observer* newspapers work fine through boards of trustees controlling their destinies. Licences have been allotted to a few community radio stations. The BBC offers a public service model for control. Is this enough? Is this the right sort of control?

Allied to this, within the main issue of who controls and how, is the further question of the kinds of check on control. I have already pointed out that in general the media are self-censoring. But are there enough checks on their freedom of action? Are the checks themselves undesirable? For instance, is the ITC a strong enough body to monitor programming, or is it really helpless in the face of the power of the big companies to dominate the whole commercial system through their control of product? How can satellite television be effectively controlled, given the fact that programmes can now be picked up which originate from outside Britain? But then again, are the checks on control and ownership actually too strong? The Government licenses all broadcasting, and some people say that it has allowed too few licences too slowly where local and community radio is concerned.

2.3 Access

Who should have access to the media, why and when? The Home Secretary has the right to demand air time any time he pleases if he judges it is necessary in the public interest. Those who are defamed in any way by the press or broadcasters have no right to anything – no apology, no space. We have regional and local television and radio stations. Commercial stations are awarded contracts on the understanding that they will provide local programming. But local people do not have a say in this. They do not have a say in what committees are set up to advise the owners. They certainly don't have a right to any air time, any more than they have a right to space in what is often proclaimed to be 'your newspaper'.

The question of access for political purposes is a very difficult part of this debate. Small parties frequently complain that they cannot get enough air time. The amount of air time for Party Political Broadcasts around general elections is fixed by the broadcasters and the Whips of the Labour and Conservative parties. Smaller parties such as the Liberal Democrats are involved in discussions, but they have no legal right to any particular amount of air time. The problem remains of who is to decide who gets on to say what. There is no right of access for local political views either. At least we can see and read about national political events and issues. But matters of local government are not talked about as of right – only when the local editor decides there is some mileage in them. So far as newspapers are concerned, one special problem has to do with distribution. You could produce a local paper as a way of getting access to the audience (whether to present political views or anything else). But you might not be able to distribute it because distribution is largely a monopoly of WH Smith, John Menzies and Surridge Dawson. They could decide not to handle your newspaper if they thought it was too political. This would mean it wouldn't get into their shops either.

2.4 Invasion of Privacy

How far should the media be allowed to go in respect of entering and digging into people's lives?

There has been much talk in recent years about creating a new Act of Parliament to curb the activities of the news operations of the media. There was a particularly notorious instance when pictures were printed of Princess Diana working out in a private gym. Newspapers claimed it was a matter of public interest. Certainly it would be difficult to construct an Act which did not then protect corrupt individuals from having their sins exposed. Equally, there is well-founded criticism of the press in particular for indulging in what is called cheque-book journalism in order to bribe people to expose the private lives of the famous, or perhaps even to exploit their connections with those who are victims of tragedy.

2.5 Freedom of Information

How much freedom should the media have to obtain and communicate information?

The debate here is a sharp one because unlike many other countries Britain does not have a freedom of information act. Organizations can be very obstructive about releasing information which might be of interest to the public, and there is no easy means of compelling them to do so. Even where, for example, there is a clause in local government acts which insists that people have a right to see minutes of council meetings, various councils resort to devices such as charging for the service and demanding that the enquirer (a local reporter?) has to say which pages they want to see or have copied before they will be released. This of course is an impossible demand.

These are only a few issues briefly outlined. But they are important ones because they raise essential questions about who runs the means of communication called the media, how, for whom, and with what effect.

It is clear that there are many messages that we receive through the media. These messages are often indirect and covert: there may be more in your film or magazine than appears on the surface. What the audience makes of such messages is yet another matter. Audience is always an essential factor in the process of communication. In the case of the media as with all communication, the nature of the audience affects how communication is carried on and how it is understood. Media operators are well aware of this. They shape their material with a strong sense of what will attract and please the audience.

3 MEDIA ANALYSIS

3.1 Construction and Deconstruction

A first and obvious point to make is that you should be uncovering the methods by which meaning is constructed into media material. This is done by deconstructing that same material – taking it apart to see how it works and how it came to be. Possible methods for carrying out that deconstruction have already been explained.

Image analysis can help you deconstruct meanings, for example, by awareness of the use of foreground in the frame or of the use of camera position to direct attention. Look, for instance, at street posters and see what object is in the foreground or what the camera was focused on as the shot was taken. Why is that object there? Is it important? In what way? What message does its existence and placing suggest? In some cases, you may find that it is actually the product which is there, though this is not obvious at a glance.

Content analysis can also help deconstruction. Take action comics, for example. Are they about action? Count how many frames, relative to the whole, show action taking place. Is it many, proportionally? How many stories are there, and do they have common themes of violence? Who commits the violence? Is there a message here? Are there certain kinds of repeated frame view or 'shots'? What are they? Are there some which repeatedly show close-ups of the hero's face, for example? If so, what is the message here?

3.2 Knowledge and Power

The ability to deconstruct material gives us at least some knowledge of what is going on, of what possible meanings there are, which we are taking in, one way or another. This knowledge gives power, the power to make choices about what may be meant, the power to accept or reject what we are told from a position of some strength and not from a position of weak ignorance. The knowledge is based not so much on facts as on skills – the skills of analysis and interpretation. The power lies in understanding communication properly, in gaining some measure of control over the messages we receive, in balancing the power that producers have to construct attractive and influential communication with the power of the audience to accept or reject that communication. This is not to suggest that we can come to know everything or that there is some finite truth about the media to be discovered.

3.3 Enjoyment and Pleasure

The recognition and deconstruction of messages in the media should also bring some enjoyment and pleasure. It should certainly not detract from it. Students have been heard to say that they just want to 'enjoy' their films or their magazines, and don't want to analyse them and talk about them. Being asked to think about what we read and view is not an attack on the pleasure of the experience. It should enhance that pleasure. When you look for messages in material you should get more pleasure from more understanding. There is fun in finding out and satisfaction in seeing more than one did at first reading. You need to find out what messages you may have been getting and how exactly they are getting to you. This is another kind of pleasure in addition to a first reading of the media text.

REVIEW

You should have learned the following things about messages from this chapter.

1 ABOUT MESSAGES

1.1 The meaning(s) of a piece of communication are also its messages to us.

1.2 Media material always gives us more than one message.

1.3 Common messages that we receive through the media are about lifestyle, group identity, self-presentation and our beliefs and values.

1.4 Media messages are presented in such a way that it is assumed that their meanings are 'natural' and correct.

1.5 Messages may not appear in the media by intention. But even if they are covert and unintentional one should still be aware of them and their importance.

1.6 Messages may be framed in such a way that the audience is caused to prefer one meaning or reading above all others.

1.7 Messages accumulate their effects and impact by being repeated over a period of time.

1.8 There are dominant categories of messages which may be described as those about: issues, normality and deviancy, culture and subculture, gender and age, power, ideology. They also reveal the workings of hegemony and discourse.

1.9 Some messages that we receive through the media contradict one another. Contradictions reveal the strains within the ideology.

1.10 The audience has very limited access to the media that may allow it to create its own messages or modify those represented by the media producers.

2 ISSUES

There are a variety of issues concerning the operation of the media which are sometimes also discussed in the media. Major examples are those of bias in media material, who has control over media production and message making, how the audience may have access to the media, how far the media should invade people's privacy, how free the media should be to use information.

3 DECONSTRUCTION

When deconstructing media material for its meanings one should recognize three points about the importance of media study and analysis. That the methods of study already described can be used to deconstruct the material in order to construct the meanings referred to in terms of messages. That this gives one some kind of power over the media to balance the power and potential influence of the producers. That there should be a kind of pleasure in doing this.

Activity (7)

This activity is concerned with the meanings which the media collectively create for us about our lives and our beliefs. It focuses on the idea of lifestyles, not least because the media themselves do this. Media define us by categorizing the life we lead. But in doing this they actually help define the life we lead, and create those categories. It is a bit like living in a hall of mirrors facing one another. You need to recognize that the mirrors are there, to stand away from them in order to understand what they are doing to us. On the surface, lifestyle is about trappings – the products which some parts of the media are actually selling us. But it is also about the beliefs and values which go with the lifestyle. This activity is more descriptive and analytical than some previous ones.

Lifestyles and New Worlds

To carry out these activities you will need one or more general mail-order catalogues, and magazines concerned with the home, such as *Homes and Gardens.* From these sources you are asked to –

ANALYSE THE MAGAZINES for the different kinds of lifestyle which are represented – i.e. those which centre on posh homes, or on beach holidays (see also the list below).
CREATE A COLLAGE for one or two of the lifestyles.
WRITE ABOUT the lifestyle and your collage.

For the collage, it is suggested that you could represent the following lifestyle topics: the rooms of a house; the garden; holidays; leisure activities; possessions. These items will add up to the external signs of the lifestyle of those who are supposed to inhabit this media world.

For the writing, it is suggested that you could deal with questions such as:

- What are the beliefs and values shown?
- What objects are valued above others, and what does this signify?
- Who has what kind of power in these worlds? E.g. men's power may be about DIY tools, women's power may be over the bathroom!
- How do these worlds differ from your own world and lifestyle?

Ultimately, this activity will be leading you back to ideology. Not only are the invented lifestyles an expression of power over the consumer, an attempt to put us in a position where we value these worlds and all their works. But also within the lifestyles you are likely to find all the values and beliefs of our culture and our ideology. The most obvious values will be to do with gender. But close behind will come meanings about the value of consumption and then, probably, implicit statements about ideas such as class.

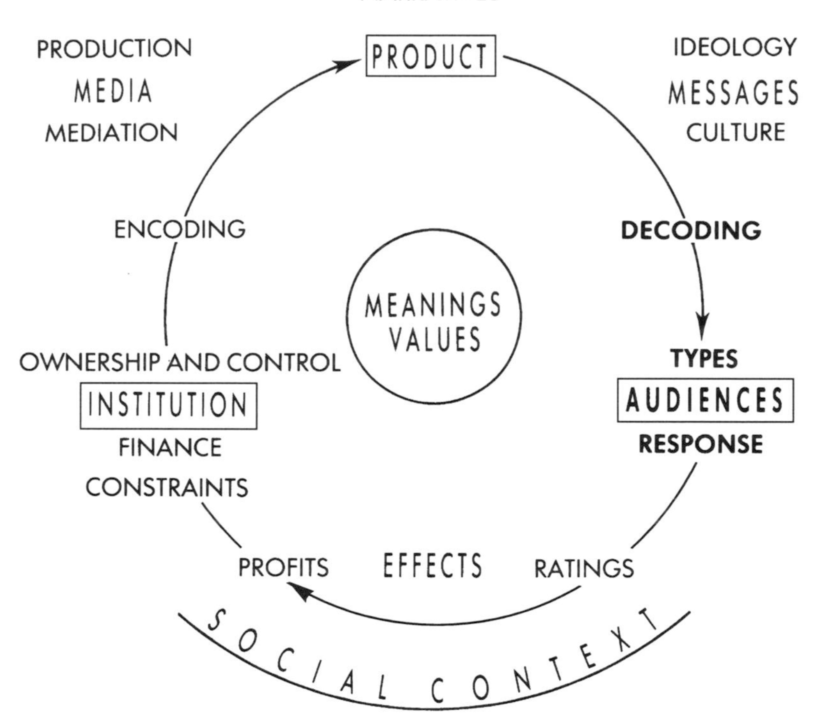
REPRESENTATION
INFORMATION AND PERSUASION
GENRES
REALISM
NARRATIVES
PRODUCTION
MEDIA
MEDIATION
PRODUCT
IDEOLOGY
MESSAGES
CULTURE
ENCODING
DECODING
MEANINGS
VALUES
OWNERSHIP AND CONTROL
INSTITUTION
FINANCE
CONSTRAINTS
TYPES
AUDIENCES
RESPONSE
PROFITS
EFFECTS
RATINGS
SOCIAL CONTEXT

8

Audiences

We have reached the point in the communication process where we are dealing with those who are both the receivers of messages and the ones who pay for it all, however indirectly.

In terms of mass media, the question of who the audience is and what effect they have on communication is a complex one because of the size and variety of audiences involved. Perhaps the first thing to realize is that the **audience is not something separate from** us. It is not some abstract idea. We are the audience – or different audiences. **The people who make media material are at the same time part of the audiences for their own material.** Press editorial teams also read other newspapers – avidly. Music A & R people also listen to records, and so on.

At the same time, it is silly to pretend that these people are just the average person in the street. In terms of their skills and their position they are also special. People who edit film and television have a special power to shape the material which we spend a great portion of our lives taking in. They are in unusual jobs. There are not many of them. They have special skills. So to this extent they do have a special place in the media communication process. It has been remarked on that they are also special in other ways – for example, they are predominantly white, male and middle-class. Those in senior positions of power are also dominantly middle-aged with the background of a university education. To this extent, while actually part of the audience, they are not like the majority of that audience.

So we need to start looking at audience by taking on two opposing yet equally valid truths. **In one sense there isn't a gap between producer and audience, and we shouldn't think in terms of them and us. But on the other hand 'they' actually are rather separate from the audience in general by virtue of background, skills and their very power to produce these pieces of communication.**

1 DEFINITIONS OF AUDIENCE

In general **we can describe the audience in terms of scale and of specificity.** Scale relates to what is loosely called the **mass audience.** This mass audience is a particular phenomenon of the media and of the twentieth century especially – huge numbers of people all reading or viewing the same product. The significance of this lies in the possibility that so many people might make the same reading of the same material and end up with the same views. Of course they might *not* take the same meaning from the material, and they might *not* do anything specific as a result of taking in those views. Nevertheless, we cannot ignore the possibility.

The audience figures in Figure 8.1 look generally impressive. Those that run into millions stand for an enormous consumption of messages. Those from broadcasting represent a simultaneous consumption – a sort of mass behaviour which the electricity and water authorities know about when enormous numbers of people flush the toilet and make tea in the advertising breaks. Figures like these look significant on grounds of sheer scale. How far they really are significant has to be determined. But at least it is clear that, without needing to argue whether 100,000 or 20 million is large, **media audiences are not only huge but distinctive because they are huge.**

We will look at potential effects later on. But it is worth saying now that this matters because of the possibility of informing or persuading or influencing such a large proportion of the population at one time. It matters because, given this scale we can reasonably talk about the media as a socializing influence. That sort of proposition would not make sense if, as in 1937, there were only 100 television sets in the country (in London). It matters because **economies of scale mean that there are huge profits to be made from these mass audiences.** The media have become colossal industries, as vital to our economy as manufacturing industries. As mass audience we not only put millions of pounds in the pockets of media owners, but we also indirectly pay the wages of thousands of our fellow citizens who work in those industries, making

Television	Average peak programme viewing figures	18 million
Radio	Average peak programme listening figures	4 million
Daily newspapers	Average total sales figures	16 million a day
Cinema	Average admissions	2½ million a week
Records/tapes/discs	Average sales figures × units sold	3⅓ million a week
Books	Average sales figures × units sold	10 million a week

Fig. 8.1 The consumption of British media: typical audience figures

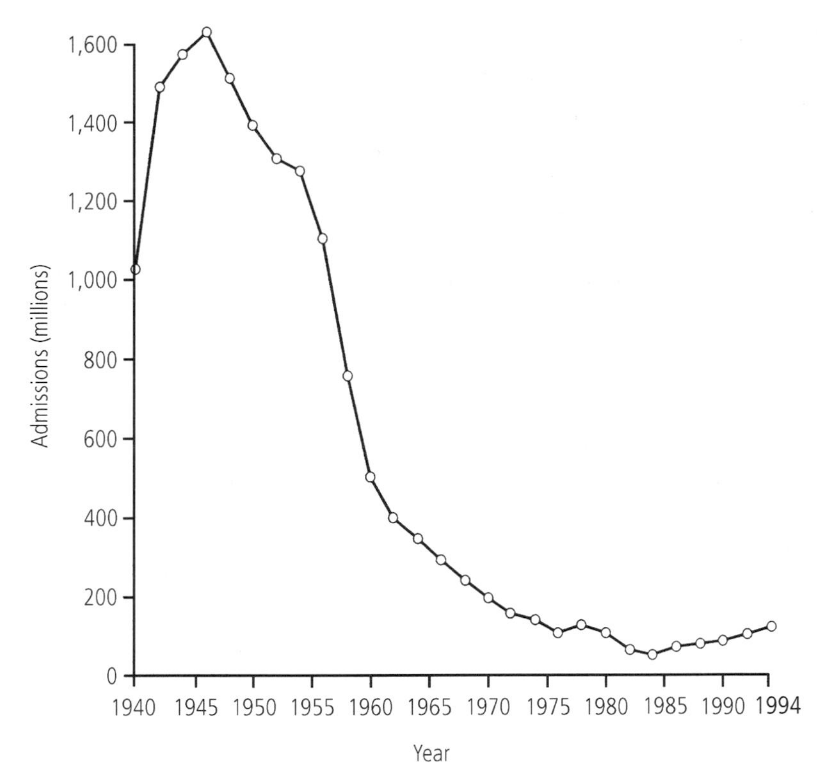

Source: Guiness Book of Film Facts and Feats & Cinema Advertising Association.

Fig. 8.2 Cinema attendances, 1945–94

Figure 8.2 shows how the cinema audience has collapsed over the last 50 years. But how many other ways are there for people to see films?

equipment, making programmes, making communication artifacts such as tapes and discs.

Yet having said all this, I would suggest that it is the definition of audience in specific terms which really matters, especially in arguing the case for any kind of influence.

2 SPECIFIC AUDIENCES

These audiences in their various categories still involve vast numbers of consumers. It is important to realize that there are few examples of mass communication which address a real cross-section of the population by age or background or any other criterion. The nearest any of the media get to this is the 23 million who watched the popular sitcom *Only Fools and*

Horses at Christmas 1996. **Television has many different audiences for different programmes.** The same people do not consistently watch the same programmes – even, say, a drama series. On the other hand, there are similar types of audience for given programmes.

Audiences are specific in three complementary ways:

A **They are defined by the particular magazine, record, film which they consume.** So we talk about the audience for *Time* magazine or the audience for records by Oasis.

B **There is a specific audience for a type of product** – computer magazines, modern jazz music, romantic films, and the like.

C Audiences are specified by **the audience profile**, in terms of standard factors such as age, class, gender, income, lifestyle, and so on.

So there is a very specific audience for a given magazine for women. There is a specific audience for women's magazines in general. There may be a describable audience for media material thought to interest women in particular. Similarly, there is an audience for a specific section of an upmarket newspaper – say, on science or books. There is a specific audience for the newspaper as a whole. There will be a type of audience which prefers to buy that type of newspaper.

We have recently had an example of the economic importance of specific audience in the explosion of European-wide magazines for women, like *Bella* and *Best*, which work to a formula and just vary some of the material for some specific national audiences. Another example is *The Independent* newspaper. This broadsheet, quality paper was started in 1991 to target a well-off, politically aware, middle-class audience which does not identify with the right wing in politics. It now has a Sunday sister paper. It has been quite successful, has profitable advertising rates, because it has found a market niche, a particular audience.

All of this adds up to the idea that, even on a mass scale, you have to have a sense of who you are communicating with and why in order for that communication to be successful. Sheer size of audience alone is not enough for media producers to succeed. The concept of audience is important at least in respect of defining the material and helping it to sell.

The quality of what is said to those audiences is another matter, as is the meaning and value of the messages delivered. It is the idea of audience which we must recognize if we are to understand how the media operate and how we get the kind of media that we do. Any notion that the producers are crudely spraying messages and material at us on a take-it-or-leave-it basis is naïve. They are very well aware of the need to target and recognize audiences, even to build audiences by shaping material. There may be no altruism in their reasons for doing this. But there is some truth about effective communication in all this. And you as students of the media need to be able to identify the characteristics of various types of audience (as well as how they use material) in order to be able to understand how communication takes place. These ideas of targeting and of using material actively are the subject of the rest of this chapter. So now read on.

3 PRODUCT AND AUDIENCE

It is a principle of communication that the way one says something, what one chooses to say, is shaped for the receiver. To this extent communication must have something of the receiver in it. But the volume of material and the range of forms in the media, allow us to see more of that audience in the communication. To make an obvious point, photoplay magazines for young females almost literally represent their audience in the examples chosen to be photographed in the stories. There is no doubt who this communication is for. This kind of identification helps the audience relate to the product. You can get hold of these magazines, and see what other points of identification there are. For example, what is said in such stories about lifestyle, values and dreams? Should we believe the audience has these in their heads and is therefore attracted to them in the product?

The media have considerable financial and skills resources, as well as time and experience with which they can make their words and pictures recognize and define the audience. This last point is important. By saying things that are relevant to a particular audience, **by bringing that audience into the material, the product actually defines the audience for whom the communication is intended.**

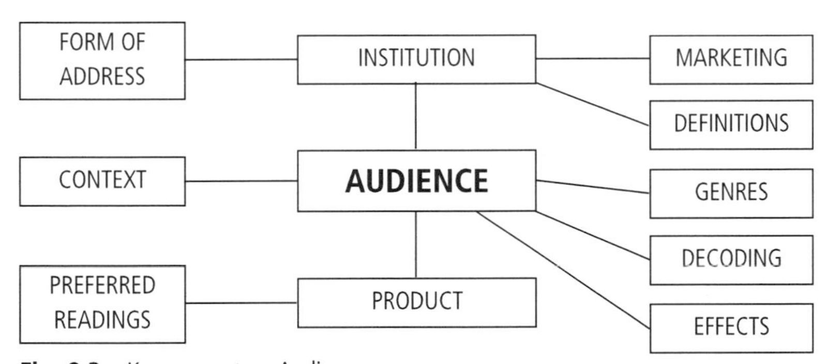

Fig. 8.3 Key concepts – Audience

Heavy metal magazines contain photographs of clothes, of fans, of performers which define their largely male audience in the product and by providing a model for lifestyle which is assimilated into the behaviour and habits of that audience. They are shaping the material for an audience which they have already shaped through previous material. Obviously there are other inputs into the lives of heavy metal fans – concerts, group talk, and so on. But still there is a sense in which that circle is a very tight one. Ask yourself, how do I know, by reading this magazine, what kind of audience it is intended for? What do I know about this audience from the magazine? What kinds of audience is this magazine certainly *not* intended for? More formally you can ask, how is the audience represented? In that case look at elements such as appearance, lifestyle, occupation, relationships, beliefs and attitudes.

The English newspaper the *Daily Mail* is another example in which we can see the audience in the choice of typical articles and sections, in the style of address, in the editorials, as well as in the newsphotos and advertisements. It is a middle-class, middle-of-the-road paper, with a slant towards a female audience, and assumed affluence signalled by sections on money matters.

This newspaper does have an audience which can be read into the material, and which is targeted through the choice of content. The repetition of similar items and the proportion of space that they occupy suggests that their inclusion is intentional, though this may be an unintentional expression of values. Whatever, as the audience reads the paper it is arguable that it is accepting a stereotypical view of its own composition and interests.

4 MARKETING AND AUDIENCE

Marketing seeks to promote the value of a product and to sell it as profitably as possible. Mass media are marketed, usually in terms of specific books, tapes or programmes, though sometimes in terms of the medium (see the Adshel campaigns to market bus shelter advertising space). As part of this marketing, the audience themselves may be marketed.

4.1 Marketing of the Audience

In the case of television the marketing people talk about delivering so many TVRs to the potential advertiser. These Television Ratings represent the percentage of the target audience that is watching when the advertisement is run. If the percentage is low then repeats may knock up the TVRs. The number of TVRs promised can be delivered by picking the right time of day or the right area or by simply repeating the advert enough times. So what the ITV company is saying to the advertiser is: we can sell you an audience – perhaps children aged 6 to 12, or married women aged between 20 and 40. Of course they will not know that only these people are viewing at the chosen times. But their market research will have pinpointed viewing habits accurately enough for them to be sure that a significant number of the target audience will be viewing, and can be 'delivered' (like produce) to the advertiser.

In fact the audience (or audiences) become a product to be marketed to advertisers in all the mass media. And advertisers nowadays do not just want mere quantity of audience, they also want to know what type of audience will be reading or viewing. Advertising space in magazines and newspapers can only be sold if the people paying for it believe that they are getting to a target audience, not just any old audience. That audience may be described in some detail. Media marketing executives can describe the buying habits, social habits and most importantly average disposable income of typical members of their target audience to the potential advertiser. They can describe them according to the government classifications (see Figure 8.4), through which we are all rated in terms of types of occupation (and therefore income).

Group	Description		Percentage
A	Upper middle	Business/professional Upper management	3
B	Middle	Professional Middle management	13
C[1]	Lower middle	Supervisory Trades	22
C[2]	Skilled	Blue collar Skilled manual workers	33
D	Semi-/Unskilled	Manual workers	20
E	Subsistence	Pensioners etc.	9

Fig. 8.4 Socio-economic groupings: a chart describing the audience by socio-economic grades with percentage of newspaper readership held

4.2 Marketing to the Audience

The audience is recognized as a factor in the way that it is sold the product, as well as in the way that the product itself is shaped. The best example is again television, where **programming and scheduling are used to define broad types of audience**. For example there is a peak time audience from 7.30 to 10.30 in the evening. Children's viewing runs up to 9.00 (in theory!), but there is a crucial audience of young people from 4.00 to 6.00. The daytime audience previous to this is dominated by elderly people and by women looking after children – for obvious reasons. Then there is the late night audience, which has a greater proportion of males than earlier in the evening. The Saturday night audience is smaller because of leisure habits (people go out!) and so the costs on the advertising rate cards go down slightly. There is the audience for religious programmes on Sundays. There is the child audience during school holidays. There are the mass family audiences during Bank Holiday times, especially Christmas.

These are defined and recognized, and are being sold to through programming and scheduling. What this means is simply that certain types of programme are put out at certain times for certain **audiences.** You can prove this by going through the TV schedules over a week, looking at the times and the types of programme. If you do this, you will find yet more types of audience that are identified, for whom the programmes are being marketed (such as the weekend audience for sport). In fact you should look at the schedules and ask of any programme, why is it placed here?

None of these programmes appears by accident. Senior executives are paid a great deal of money to **programme material** so that it will attract the right audience at the right time. They use devices like putting out generally popular soaps early in the evening to pull in the viewers, since there is some evidence that people are lazy about switching channels. They 'hammock' a 'weak' programme such as a documentary between popular ones such as a comedy and the main news. They try to hold the audience for the whole evening by having two or three popular comedies or dramas in the first half. They try to pull the audience back the next evening by having a 'blockbuster' spread over two evenings. And of course they market directly by having attractive previews for the evening or the week repeated between programmes.

4.3 Targeting the Audience

Essentially this notion is about first actually identifying and describing the target audience, and then about shaping the marketing strategy and materials to the lifestyle and needs of that audience.

There is a Target Group Index produced by the British Market Research Bureau, which interviews thousands of people in order obtain information about their social behaviour and habits, as well as of course their preferences for certain products. From this information it is then possible to devise categories or descriptions of target groups, which may have nothing to do with conventional socio-economic descriptors. Some of the slickest descriptions are contained in phrases like 'the pink pound' or 'Dinkies' for double income, no kids.

Targeting is most obvious when one sees different adverts for the same product. The advertiser is trying to match the attitudes and values of the target audience. Vestergaard and Schroder talk about a 'signification process whereby a certain commodity is made the expression of a certain content (the lifestyle and values)'. The advertisement is trying to link values to the product and vice versa, so that in the minds of the target audience the product is the lifestyle and has value.

4.3 Marketing to Find an Audience

Sometimes **attempts are made to use marketing as well as new products in order to find and define an audience of a type which did not exist before.** Of course we are not talking about suddenly finding a section of the population who never read magazines but finding a section who will take to a new magazine because it offers things that others do not. When *Cosmopolitan* first came out it was sold, like *Options* after it, to a 'new' kind of woman. It recognized, implicitly, social changes and attitudinal changes. These meant that there was a female audience for a magazine which at least varied the formula for women's magazines enough to take account of independence of thought, of lifestyle, and of income. The product was marketed to find an audience.

Another famous example was the 'discovery' of the television audiences for snooker. The general principle of good audiences for sport is well established.

So once again, we are not talking about some dramatic feat of marketing and audience discovery. But the BBC did identify a potential audience, offered the product to it, and then, more positively, sold that product, especially when it was discovered to be really successful.

4.4 Marketing Across the Media

This is akin to discovering and building variations on the audience by selling basically the same product to an ever-expanding audience across different media. For example, Disney movies such as *Aladdin* or *Pocahontas* are certainly not sold to one audience just as a film. The soundtrack albums were also sold to that audience and to a wider one (people who have not seen the films). And again, the videos were sold to a wider audience beyond those who had seen the films. After this there is the matter of selling spin-off products. Marketing across the media can sell more products to a greater number of the same audience. It is also possible to pick up different audiences in this way.

Another longer running example is Sue Townsend's *The Secret Diary of Adrian Mole*, best known through the two books which have sold over 6 million copies. But in fact the story started as a short play for the theatre, then became a short radio play, then became the first of the books. It has since been made into a longer theatre play and a television series. In other words it has crossed over the audiences for the various media concerned. It has expanded its audience in terms of number and profile, moving from something relatively cultish for a smaller audience to communication which has mass appeal.

5 MODES OF ADDRESS

In this case one is looking at something like style, at **the way in which the media text 'talks to' the audience verbally or visually**. This is all about making a connection, about being attractive to that audience, about defining the relationship of the text to the audience.

For example, television programmes for young people such as *Live and Kicking* on a Saturday morning, talk to their audience in a very particular (and peculiar?) way. The presenters talk very fast and move around a lot and fool around as if they were hyperactive teenagers – in fact they are nearly 30 years old. They appear to be generating the persona of an ideal elder brother or sister for the young audience. This mode of verbal and non-verbal address is supplemented by the use of visual address, which is characterized by rapid editing, bright colours and brief bites of items. The mobile camera and changing camera positions create excitement, but are anchored by the presenters' determined direct address to the audience out there.

This mode of address is used, with variations, on most so-called 'youth' programmes. It is meant to appeal to the audience, to create identification with that audience. It is also about creating an identity for the genre. This identity

girls' night

You don't have to bust the elastic in your disco knickers to have a brilliant night – in fact, you don't have to leave the house! Follow Sugar's guide and we guarantee the perfect night in.

Evenings in on your own can be totally chill-some, but if you want to make a proper night of it, invite some chums round (preferably pals you feel really relaxed with) and get them to bring some of the following:

time-out togs!

This is a night of total luxury and indulgence, so throw off your day clothes and anything restricting, and snuggle into some cosy, comfy togs:

- Big, old, sloppy T-shirt and tracky bums.
- Leggings with baggy knees.
- Dad's old holey cardis and jumpers.
- Dad's socks; there's nothing fabber than snuggling in his chunky-knit foot holders!

win!

THE ULTIMATE TIME-OUT TOGS ARE THESE RED AND WHITE, CHECK HEART PJ'S (KNICKERBOX, £35)

girlie chats

There's nowt better than sisterly bonding and getting down to some serious talking with your pals. Stuff to try:

- List the top 100 sexy boys in the world.
- Describe dream dates to one another.
- Discuss what you'd call your children. Cringey, but we all do it!
- Laugh about the old days and discuss how you've changed.
- Do Sugar's quizzes on each other.

chill-out chow

Before you veg out completely, fill your fridge with scrumptious snacks to munch on throughout your chill-out mission:

- Häagen-Dazs Cookies and Cream – eaten straight out of the tub!
- Pot of Tea and choccy Hob-Nobs (warning: do not dunk for longer than 3.5 seconds!).
- Mountains and mountains of popcorn – salted, buttered, toffee...
- Make a Victoria sandwich cake in the microwave. Yup, these delicacies are available from your local supermarket!

CHOCCIES ARE ESSENTIAL AND OUR CHUMS AT CADBURY'S CREME EGG ARE GIVING AWAY TONS OF 'EM (PLUS! RONAN FROM BOYZONE AND RYAN GIGGS LIFESIZE CUT-OUTS)!

win!

super sounds

No perfect night in is complete without some funky tunes in the background:

- *The Best of All Woman* – fab diva album.
- Massive Attack's *Protection* – cool vibes.
- *Serve Chilled* – funky tunes for chilling.
- Boyzone's *Coming Home* – out March 5, a new, special limited edition single.

ALL THESE TASTY TUNES ARE UP FOR GRABS!

win!

wondrous watchings

Armed with your nibbles, dump some duvets and cushions in front of the telly, turn the lights down and stick on some groovy movies! Make sure there's a box of tissues close at hand for the soppy bits

gotta-see-'em-again flicks:

The films you'll want to rewind and watch over and over:

- ***Point Break*** Daredevil action movie. **Watch in slow-mo:** Keanu's bum flashes.
- ***Interview With The Vampire*** Spooky tale from the crypt. **Cushion-diving:** Brad "necking" with a victim on the sofa.
- ***Thelma and Louise*** Ultimate chick flick. **Watch it 100 times and you'll still blub at:** the chick chums' cliff-top decision.
- ***Before Sunrise*** continental love story. **Sigh "bless" at:** the will they?/won't they? snog-up moment in the sound-booth!

58 Sugar

Fig. 8.5 Girls' night in

This page from a teenage girls' magazine typifies the formula in terms of reference to food, clothes, boys, music, TV, and films. It also illustrates how media product tries to 'address' its audience through words and pictures. In this case, it is easy to spot the chatty big-sister style which addresses the audience in familiar tones and a special language – 'comfy togs', 'veg out', 'cool vibes' – and other recycled terms of earlier decades.

and its characteristics can even be seen in other media – chatty, bouncy, colourful, 'noisy' magazines for teenage girls are an example.

In one sense this sounds all very well as a way of appealing to the audience – talking to them 'on their level'. However, Gill Branston points out in *The Media Studies Book* (1991) that one could be looking at the idea of audience address the wrong way round. Perhaps teenage girls respond to targeted magazines and programmes beause they have become conditioned to respond in this way. Perhaps media material which repeatedly assumes that its audience has the attention span of a gnat, ends up convincing the audience that this is all it is capable of. Another negative aspect of mode of address would be when an audience member feels disturbed because they perceive the material as being aimed at them, but the person being addressed is not really like them at all. What about the teenager who actually feels guilty or alienated because she knows that she is not like the female that the magazine is talking to, but believes that perhaps she ought to be?

All examples of media address the audience in some way. Some modes of address have a more distinctive identity than others. It is relatively easy to pick out the authoritative, firm voice of broadcast news, for instance.

But, to take one more example, modes of address permeate advertisements. In fact if one turns the pages of a magazine or watches a sequence of TV commercials, then one is addressed in different ways through different 'voices'. Some adverts are intimate and whisper in our ear; some are strident and assertive. Some talk to us very reasonably about why we should use the services of a certain insurance company; some talk to us as if we are concerned parents. And this last point draws attention to the fact that mode of address does position the audience in relation to the text – 'we, as audience members, are responsible for making sense of the media text, but it also creates for us particular positions from which to understand it' (Branston, 1991).

We are addressed as if we are a certain kind of person, as if we have a certain kind of role. This positioning helps set up a preferred reading, and it takes away at least some of our freedom to read the text as we please. It is all part of the persuasive manipulation of advertising in this case.

6 ACTIVE AUDIENCES

We have looked at various ways in which audiences may be defined, including through the media product, and at ways in which they are addressed through marketing of product. But now we must dispel any notion that audiences are simply passive receivers, that they are lumps of clay to be dug out and moulded. Although their scope for providing influential feedback on production itself is very limited, in terms of doing something with the communication that they receive, audiences are not as slothful as the mythology of the couch potato might suggest.

Kilborn (1992) in talking about soaps, refers to the ways that viewers tap into 'a whole range of reading skills that they have acquired over a period of time'. He argues that soaps in particular encourage social discussion and reflection because there are different ways in which one can make sense of their narrative.

6.1 Processing Information

For a start, audiences are receivers of messages, they are processors of information. **They may be sitting and reading or viewing in a passive sense physically. But in fact a great deal is going on in the mind.** And people do many things while keeping an eye on the screen, including talking about other things!

The stream of information which comes from the pictures and print on a newspaper page requires active involvement to decode it. Reading is an activity. It requires us to recognize symbols, to make sense of them, to assign meaning to them. We assimilate those meanings into our store of knowledge, into our opinions, into our views of the world.

In another sense we are always learning from the media when we process this information. Examples such as *Sesame Street* make it most obvious that this learning is taking place.

Sometimes the activity involved becomes externalized so that it is very clear just how much is going on inside our heads. Cinema audiences can become so active (involved) that they actually throw comments at the screen, and make noises of protest or approval.

6.2 Indirect Active Response

Another kind of audience activity which is not immediate but which clearly shows a response to the communication, is the increase in participation rates in certain sports after exposure in the media. Obviously, media sports programmes may not be the only factors causing people to join sports clubs in greater numbers. But sometimes they seem to have been the major factor – as in the case of snooker and the rise of local halls and clubs in the late 70s. American football is a more recent example, where the development of active club interest has coincided with the featuring of the game first on Channel 4, and then on satellite channels.

6.3 Uses and Gratifications Theory

This theory says that, far from being passive consumers of media material, **the audience actually uses media material to gratify certain needs that it has.** Viewers choose to watch a certain programme because it satisfies particular needs that they have at that time – perhaps for information. Or readers select from a newspaper turning past the hard news to a gossip section on the aristocracy or stars because they need entertainment. Needs also colour the interpretation of material. One person may understand their favourite soap opera in terms of conflict because they are relating their own needs for power to what is happening to the characters. But another person may understand the same programme as being mainly about love and understanding because they

	Often %	Occasionally %	Rarely %	Never %
I watch to see a specific programme that I enjoy very much	77	18	3	1
I watch to see a special programme that I have heard a lot about	41	48	9	1
I watch just because it is a pleasant way to spend an evening	34	39	17	10
I watch just because I feel like watching television	26	42	20	13
I watch because I think I can learn something	21	46	22	10
I watch because there is nothing better to do at the time	15	32	32	20
I turn on the set just for company	14	26	27	32
I start watching one programme and then find myself watching for the rest of the evening	13	36	33	18
I watch as an escape from everyday concerns	13	27	32	27
I start watching because someone else is watching and seems to be interested	11	42	29	17
I watch to be sociable when others are watching	11	34	29	25
I watch in case I'm missing something good	11	30	34	25
I watch just for 'background' whilst I'm doing something else	10	29	30	32
I watch a programme because everyone I know is watching	7	24	38	31
I keep watching to put off doing something else I should do	7	21	33	37
I watch to ignore or get away from people around me	4	11	30	54

Base: All TV viewers. Note: 'Don't knows' have been excluded. 'When you watch TV, how often does each of these reasons apply?'

Fig. 8.6 Why do people watch television?

Figure 8.6 suggests that there is a fair degree of intentionality in TV viewing. Audiences actively make plans and decisions about their viewing selection.

have strong needs for security and choose to bring out those aspects of the story in their minds.

The uses and gratifications theory has been discussed by writers such as Dennis McQuail (*Mass Communication Theory*, 1983). In essence it suggests that we have the following kinds of need that we may gratify through kinds of viewing and reading and listening:

- **the need for information**, based on sheer curiosity as well as the practical advantages of building up a picture of the world;
- **the need to maintain a sense of personal identity**, by checking media role models for behaviour;
- **the need for social interaction**, developing a notion of one's social behaviour and relationship with others by using examples from the media;
- **the need to be entertained and diverted**, escape from immediate anxieties or achieving kinds of pleasure.

This theory relates ideas about people and their motivation to ideas about mass media. It relates interpersonal and mass communication. It helps integrate **an essential principle of communication – that we are all driven by communication needs in any area of our communication activity.**

So from this point of view the audiences for various media communication are indeed active users. They don't just take anything that they are given – though they can only choose from what they are offered. They use what is on offer to suit their needs. They select what they want from among whole programmes and newspapers and records. **They play an active part in creating meaning from the communication.** They decide what the messages may be, they interpret them.

But this theory has to be taken alongside other ideas such as preferred readings and covert messages. You should realize that media producers are well aware of ideas about audience needs (not least those who create advertisements). They are certainly capable of arousing needs and using them to sell their product, and perhaps other messages carried along in the material. So the truth, as far as we can understand it, lies somewhere in between extreme notions of audience as passive victims of cunning media producers, and audience as active controllers simply using media material as they please. And one cannot generalize about the audience anyway. Individual audience members may be more or less active or passive. Even the needs themselves are a kind of assumption, read into peoples' psychology by observing their behaviour. But then their behaviour includes media consumption, so does media product recognize a need or does it actually create it?

In so far as the audience uses the media to gratify needs, various needs can be gratified at one time. Certain tracks on a record may satisfy personal needs (for instance self-insight) as well as providing relaxation and entertainment. A need for information can be met by a fictional serial on television. At the same time, this material could also satisfy social needs, perhaps as a substitute for real-life friendship.

7 AUDIENCE, MEDIA AND CULTURE

Notions of media affecting audience or of audience using media may not be the best way of understanding the relationship between the two elements. Perhaps

they both exist in a dynamic, interactive relationship within the context of culture and the meanings produced about that culture. For example, teenage magazines for girls certainly are part of their subculture. Readers can share the culture through the magazines. They can find out who is 'in', who is 'out'. The magazines help perpetuate the argot referred to by Michael Brake in *Comparative Youth Culture* (1985), as one of three defining factors (together with 'image' and 'demeanour'). But those same female readers do have a subculture without the magazines. They go to certain clubs and pubs, they talk about certain topics, they engage with those who are selected as belonging to their subculture through social processes – not just because the magazine or record industries say this is their niche. As a consuming audience they could be defined as buyers and readers. But as an audience and subculture they exist with, but not because of, the media product. Lisa Lewis in *Television and Women's Culture* (1993), talks about 'the dynamic relationship between mediated texts and social practice'.

Lorimer asserts that 'a straightforward and comprehensive method for understanding audiences does not exist'. He talks about approaches to audience study which 'conceptualize audience/media engagement' in terms of 'elements such as gender and cultural identification'. One needs to look at how audiences behave, as much as at what audiences are. Institution, audience and text are, from this point of view, part of one thing, one phenomenon. All of these elements could be said to give or to produce meanings about our world. But they don't do this independently. What we think that love and romance really is, does not come simply from the romantic novel industry. It doesn't come only from romantic texts. But nor

United Kingdom Percentages

	4–15	16–24	25–34	35–44	45–54	55–64	65 and over	All aged 4 and over
Drama	19	28	28	26	25	23	24	25
Light entertainment	20	19	18	18	17	16	17	18
Films	13	17	16	17	17	16	12	15
Documentaries and features	7	11	13	13	14	15	15	13
News	6	6	8	10	11	13	14	11
Sport	7	10	10	10	11	12	13	11
Children's programmes	25	6	4	3	2	1	1	5
Other	3	3	3	3	2	3	4	3
All programmes	100	100	100	100	100	100	100	100

Source: Broadcasters' Audience Research Board; British Broadcasting Corporation; AGB Limited; RSMB Limited.

Fig. 8.7 Types of television programme watched: by age, 1994

United Kingdom	Hours and minutes per week
	1994
4–15	5:54
16–34	17:19
35–64	18:17
65 and over	18:07
All aged 4 and over	16:07
Reach[1] (percentages)	
Daily	67
Weekly	84

Note: 1. Percentage of the population aged 4 and over who listened to the radio for at least half a programme a day.
Source: British Broadcasting Corporation.

Fig. 8.8 Radio listening: by age, 1994

does it come independently from the minds of the audience. It is created by the interaction of all three.

8 AUDIENCE AND PLEASURE

One proposition about this interaction is that it depends on kinds of pleasure where the audience are concerned. Critics have talked about the **pleasure given by the text.** They have discussed **pleasures obtained from the text** – not necessarily the same thing. This pleasure idea is not far from that of uses and gratifications. For example one could talk about the pleasure of recognition and identification which women may obtain from the experience of soap opera. There has been much discussion of male pleasure in looking where pornographic magazines or licentious film scenes are concerned. The pleasure is something that involves the audience in the text, perhaps binds the audience to the text, especially in respect of people being for example 'glued' to a series on television. The meaning and nature of pleasure is complex. For example, if the audience enjoys a happy ending where a love relationship is confirmed, is the pleasure one of emotional warmth in the shared and approved of experience? Or is it a more intellectual satisfaction that comes from knowing one had predicted this outcome, and now the plot seems to be neatly tied up?

John Fiske in *Media Texts* (1993) writes about various aspects of this pleasure. In respect of my last point, he refers to Barthes' work and the idea that we get pleasure from playing with the text. He refers to the pleasure of voyeurism in the visual media, made sense of through psychoanalysis. He talks about the pleasure of experiencing texts which both confirm rules, but which also allow the audience to safely challenge rules. This idea of 'safety' may contain a key point about pleasure and text. Whatever the pleasure achieved

Daily newspapers – Popular	
The Express	1,257,880
Daily Mail	2,049,100
The Mirror	2,484,238
Daily Star	668,694
The Sun	4,057,668
Daily newspapers – Quality	
The Daily Telegraph	1,040,316
Financial Times	303,573
The Guardian	398,661
The Independent	281,588
The Times	669,640
TV programme magazines	
Radio Times (BBC)	1,464,392
TVTimes (ITV)	1,007,964
Women's magazines	
Best	565,388
Company	290,081
Cosmopolitan	456,394
She	252,046
Woman	800,099
Woman's Own	761,024
Woman's Realm	307,237
Woman's Weekly	754,110
Other magazines	
Here's Health	42,239
New Scientist	115,229
Computer Shopper	155,606
The Face	112,388

Source: BRAD, August 1996/Audit Bureau of Circulation.

Fig. 8.9 Newspaper and magazine circulation figures

Great Britain Percentages

	Percentage reading each paper								
	15–24	25–44	45–64	65 and over	Males	Females	All adults	Readership[2] (millions)	Readers per copy (numbers)
The Sun	30	24	20	16	26	19	22	10.1	2.5
The Mirror	15	13	15	15	16	12	14	6.5	2.6
Daily Mail	8	8	12	11	10	9	10	4.4	2.5
The Express	5	5	8	10	7	7	7	3.2	2.5
The Daily Telegraph	4	5	9	7	7	6	6	2.8	2.7
Daily Star	6	6	3	2	6	3	4	2.0	2.8
Today	4	4	4	2	5	3	4	1.7	3.0
The Times	4	3	4	3	4	3	4	1.7	2.7
The Guardian	4	4	3	1	3	2	3	1.3	3.5
The Independent	2	2	2	1	2	2	2	0.9	3.3
Financial Times	1	2	2	–	2	1	2	0.7	4.3
Any national daily newspaper[3]	58	56	63	62	64	55	60	27.2	–

Note: 1. Data are for the 12-month period ending in June 1995. 2. Defined as the average issue readership and represents the number of people who claim to have read or looked at one or more copies of a given publication during a period equal to the interval at which the publication appears. 3. Includes the above newspapers plus the Daily Record, Sporting Life *and* Racing Post.

Source: National Readership Surveys Ltd.

Fig. 8.10 Reading of daily newspapers: by age and gender, 1994–5[1]

Great Britain

	Percentage reading each magazine							Readership[2] (millions)
	AB	C1	C2	DE	Males	Females	All adults	
General magazines								
Reader's Digest	15	14	12	9	13	13	13	5.7
Radio Times	17	13	8	6	11	10	11	4.9
Sky TV Guide	9	11	13	9	13	8	10	4.8
TVTimes	7	9	10	11	9	10	9	4.3
AA Magazine	16	10	8	4	11	7	9	4.2
What's on TV	5	8	9	11	7	10	9	3.9
Viz	6	8	7	5	10	3	7	3.0
TV Quick	3	6	6	6	4	7	5	2.5
Women's magazines								
Take a Break	5	11	13	15	5	16	11	5.0
Bella	5	9	10	10	3	14	9	4.0
Woman's Own	6	9	10	10	2	15	9	4.0
M & S Magazine	14	10	6	3	4	12	8	3.6
Woman	4	7	7	8	2	11	7	3.1

Note: 1. Data are for the 12-month period ending in June 1995. 2. Defined as the average issue readership and represents the number of people who claim to have read or looked at one or more copies of a given publication during a period equal to the interval at which the publication appears.
Source: National Readership Surveys Ltd.

Fig. 8.11 Reading of the most popular magazines: by social grade and gender, 1994–5[1]

United Kingdom			Percentages
	Aged 7–10		Aged 11–14
Males		**Males**	
Beano	28	*The Sun*	22
Match	25	*Match*	18
Shoot!	22	*Shoot!*	18
Sonic the Comic	18	*The News of the World*	17
Dandy	17	*Beano*	15
Females		**Females**	
Smash Hits	16	*Just Seventeen*	41
Barbie	13	*Sugar*	39
Girl talk	13	*Big!*	34
Beano	12	*It's Bliss*	34
Live and Kicking	12	*Smash Hits*	32

1. For 7 to 10 year olds data are the percentage of children who said they read the publication; for 11 to 14 year olds data are the average issue readership.
Source: Youth TGI, BMRB International.

Fig. 8.12 The most popular magazines and newspapers read by children: by age and gender, 1995[1]

by engagement with the text, it is only a text. This not to argue that texts have no involvement with our 'real' lives, no influence. But they are after all only about representations, they are not the same thing as life. If the camera invites us to peek at the woman undressing in the window, then it is a safe experience – we could not get found out and embarrassed or prosecuted. If we gain vicarious satisfaction from the death of a villain and the restoration of moral order, then we are in no real danger from the bullets that kill him in the story. If it turns out that we are wrong about the denouement bringing the marriage of two protagonists, then it does not matter in the way that it would if these were people we really knew and really engaged with.

Audiences are an integral part of the whole process of communication through the media. In many ways they are the *raison d'être* for the media industries, because no audience means no profit means no reason for running the organization. It is the audience which makes sense of the communication. And this sense becomes all the more important because of the size of the audience, given the potential for influence, and the part the media play in the socialization of that audience.

REVIEW

You should have learned the following things about audiences from reading this chapter.

- that the people who provide media material are also members of the audience, though their position is special because they create and shape this material.

1 Media audiences are both large and specific in their characteristics. The size of such audiences is what makes them very profitable to the producers.

2 Specific audiences may be defined in terms of the product that they consume, of the type of product that they consume, of their audience profile.

3 One may 'read' the kind of audience targeted into the product that is produced for that audience.

4 MARKETING AND AUDIENCE

The audience affects the ways in which the product is marketed.

4.1 Specific audiences are sold to advertisers, and are identified through examples of material which will appeal to that audience.

4.2 Audiences are also the objects of marketing through such devices as television programming and scheduling.

4.3 Audiences are targeted by marketing strategies.

4.4 Sometimes marketing is done in order to identify a new audience.

4.5 Marketing is something which is done across the range of media. This means that the messages within the marketing campaign are reinforced all the more.

5 Modes of address, or special ways of talking to the audience, are used to identify and to attract a particular audience.

6 ACTIVE AUDIENCES

- audiences take an active part in media communication.

6.1 Part of the activity includes the decoding of the material within the mind of the audience (intrapersonally).

6.2 There is evidence of indirect active responses to the media, as when people appear to be taking up sports as a result of exposure in the media.

6.3 The uses and gratifications theory suggests that audiences make use of media material to gratify needs that they have within them. This means that the audience is not a passive receiver of communication. The essential needs that have been identified are: for information, for a sense of identity, for social interaction, for entertainment and diversion.

7 Audiences, the media and the culture which both share exist in a complex relationship in which the one affects the other.

8 Audiences get kinds of pleasure from media texts.

Activity (8)

This activity is concerned with the idea of audience. To some extent it throws up ideas about the ways in which we categorize audiences. But it is also meant to make you think about how we 'know' what we think we know about audiences, about what they like in the media, about what they take from the media, about the ways in which the media have to recognize and respond to audience preferences.

The context is the music industry, which is a tricky one to study in that information about it is not as readily available as for other media industries. Its product is distinct from the visual and verbal codes which dominate other media. And yet it is interwoven with those media through items such as promotional videos, backing music and the film sound tracks which are often marketed alongside the movie. In particular, music alongside fashion helps to define subcultures and is held to be a marker for their distinctiveness, even their value, within the main culture.

This activity has two related parts. The first part, like the figure on which it is based, is meant to remind you that there is a wide range of categories of music. People are often fiercely partisan about the type of music they like, and contemptuous of the types (and audiences) that they don't like. As a media student you need to be more detached than that, and be prepared to study and make sense of examples outside your immediate experience.

DESIGN A QUESTIONNAIRE to be administered to a wide range of people, and to investigate the following features of music in peoples' lives:

- listening preferences by types of music (use the categories suggested in the figure in the next chapter)
- listening habits, in terms of where and when people prefer to listen to music, and through what medium (i.e. radio, gigs, CDs)
- music as part of lifestyle (i.e. what activities do people like to associate with their music listening).

CONSTRUCT A MARKETING STRATEGY for a current release, based on what you have learned from the above. This would include the campaign schedule, choice of media with reasons, advertising slots chosen with reasons, devices of publicity. You could also design a CD cover, which should fit in with the campaign theme and marketing approach.

As part of your reflection on the exercise you could compare your results with those of the *Independent on Sunday* survey, and try to account for any differences (see Figure 9.1 on page 210).

You could also reflect on what *The Independent* producers can do to publicize their products, given that they don't have the money to pay for a full-blown campaign.

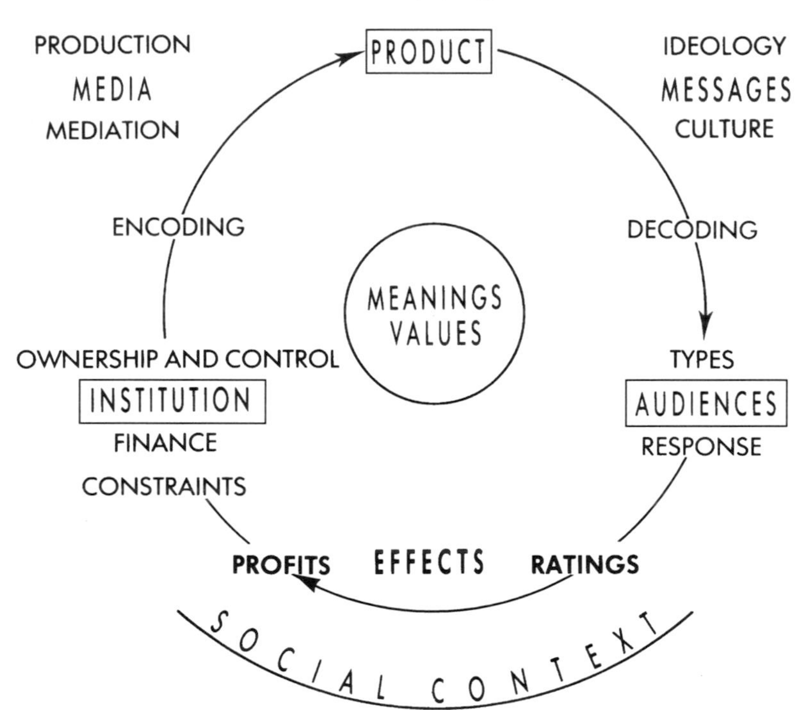
REPRESENTATION
INFORMATION AND PERSUASION
GENRES
REALISM
NARRATIVES
PRODUCTION
MEDIA
MEDIATION
PRODUCT
IDEOLOGY
MESSAGES
CULTURE
ENCODING
DECODING
MEANINGS
VALUES
OWNERSHIP AND CONTROL
INSTITUTION
FINANCE
CONSTRAINTS
TYPES
AUDIENCES
RESPONSE
PROFITS
EFFECTS
RATINGS
SOCIAL CONTEXT

9

Effects

The Payoff?

This chapter focuses on the reason why it may be thought worth studying the media – **the proposition that the media affect our beliefs, attitudes, values and behaviour, directly or indirectly.** This proposition is an expression of a belief in the power of the media.

As we have said, the media are special in various ways, not least because of the technology, the economics, the scope of their operations. They are special because they are so very conscious, in some ways, of how they carry on communication – because it pays them to be aware. They are all about a process of communication between people, in which the producers and the audience try to affect one another through their communication.

In all this, the biggest irony is that, **for all the research into** effects **it is extremely hard to prove any**. There have been many propositions made about media effects, some of which seem acceptable at least on grounds of sense and logic. A great deal has been established about conditioning factors. But there is very little which absolutely proves that if one communicates in a certain way through a particular medium of mass communication one will have a particular effect on the audience. Indeed, Curran and Seaton (1991) say 'there is no adequate vocabulary to describe the relationships between the media, individuals, and society.'

1 WHAT SORTS OF EFFECT?

There are a number of theories which provide some kind of label for effects. Then there are a number of descriptions of supposed effects which provide another set of categories.

1.1 Short-term Effects

In the early days of media research it was supposed that one could apply a simple stimulus-response model, in which the media or perhaps even a particular newspaper provided the stimulus, and the response was a change in the audience voting patterns, or something like that. This short-term behaviourist approach has been discredited. People don't act immediately, if they act at all. And there are too many variables to take account of, apart from the media.

Sometimes this approach was referred to as the hypodermic effect theory, as if people were being injected with some media material and responded accordingly. Such theorists liked to quote the famous panic in 1939 when Orson Welles radio broadcast a drama documentary version of H.G. Wells's *War of the Worlds*. Some people believed there was a Martian invasion, and thousands took to the roads. But in fact it has been demonstrated that there were other crucial factors such as people's knowledge of the imminence of real war, which had them in a state of high anxiety anyway.

1.2 Long-term Effects

What is proposed is that, whatever the nature of the effects is (see below), they take place over a long period of time, and are more to do with changes in attitude and belief than with immediate behavioural changes. This idea is plausible and underpins much existing research. There is still a problem of conducting the right kind of investigation of audiences over a long enough period of time in order to be sure that their attitudes have indeed changed. There is an even greater problem in this case of separating media influence from other influences. There is also a problem in objectifying the nature and degree of attitude change. One basic difficulty is that one has to find some external behaviour which represents the internal change in attitude. So often one comes back to asking the audience more or less direct questions and replicating experiments in order to 'get inside their heads'. One basic question then is how far one can accept or even fairly interpret what people say. What they think and what they say they think may be two different things.

1.3 Inoculation Theory

This theory proposes that continued exposure to media messages causes us to become hardened to them, to become de-sensitized. The theory was popular with those who wished to believe that the media made us insensitive to violence. The same people were not so ready to believe that it might also follow that we would become immune to persuasive devices of the media, for similar reasons. Anyway, research has found no evidence to support this theory, though, as a point of view, you will still hear it tossed around.

1.4 Two-step Flow Theory

This theory proposes that the media influence us indirectly in two stages. The first stage involves the media activity (and perhaps opinion makers): the second stage involves opinion leaders (see page 216) who are respected members of our peer groups. We listen to what they have to say more than perhaps the media themselves. If the media have influenced them in the first place, then in fact we can be indirectly influenced. This theory owes much to the work of Katz and Lazerfeld in the 1960s. One important offshoot of the work was to generate propositions about how active or passive we are as receivers of the communication. It suggested that, in talking with people around us about what the media offered, the theory indicated that we were quite active in dealing with the material, not passive sponges.

1.5 Uses and Gratifications Theory

This has been dealt with in Chapter 7 because it is all about the active audience. This assumes that there are effects but does not really engage with measurement of them. You will remember that the idea is that the audience uses media material to gratify certain needs, described generally as needs for information, entertainment, personal identity, social interaction. The theory deals with how the audience may be affected, as much as whether they are affected or not. This is not to say that the propositions have been developed in a vacuum. It is thought that people are affected within these areas because the headings classify the kind of response given to questionnaires, for example. But again, the research does not really question the existence of the needs themselves – they are there because researchers say they are there. Nor is there enough questioning of why they are there – i.e. what part do the media play in generating these needs in the first place?

1.6 Cultural Effects

Researchers here are more concerned with collective audience effects than with reactions of invididuals, with how the media define and circumscribe culture, and with the perpetuation and reinforcement of cultural divisions – stereotyping by race for example. This work also relates to views of media functions (see page 76–9).

If we now look at types of effect in terms of what may actually happen to the audience member, then you should remember that this is still only a list of categories that has been set up. There are some people who believe that these effects have been demonstrated and do happen – but not everyone.

1.7 Types of Effect: Various

Attitude Change: the media have the effect of changing people's way of looking at the world in that they modify their attitudes towards others and towards issues.

Cognitive Change: the media have the effect of changing the way that people think, the way they value things, they change (or modify) their beliefs.

Moral Panics/Collective Reactions: the media have the effect of generating unfounded anxiety about issues such as law and order or public health.

Emotional Responses/Personal Reactions: the media affect people by evoking emotional reactions (as much as rational ones). This is next door to 'moral panics'. For example, it could be argued that the question of personal and body image, especially for women, is raised emotively by the media. The

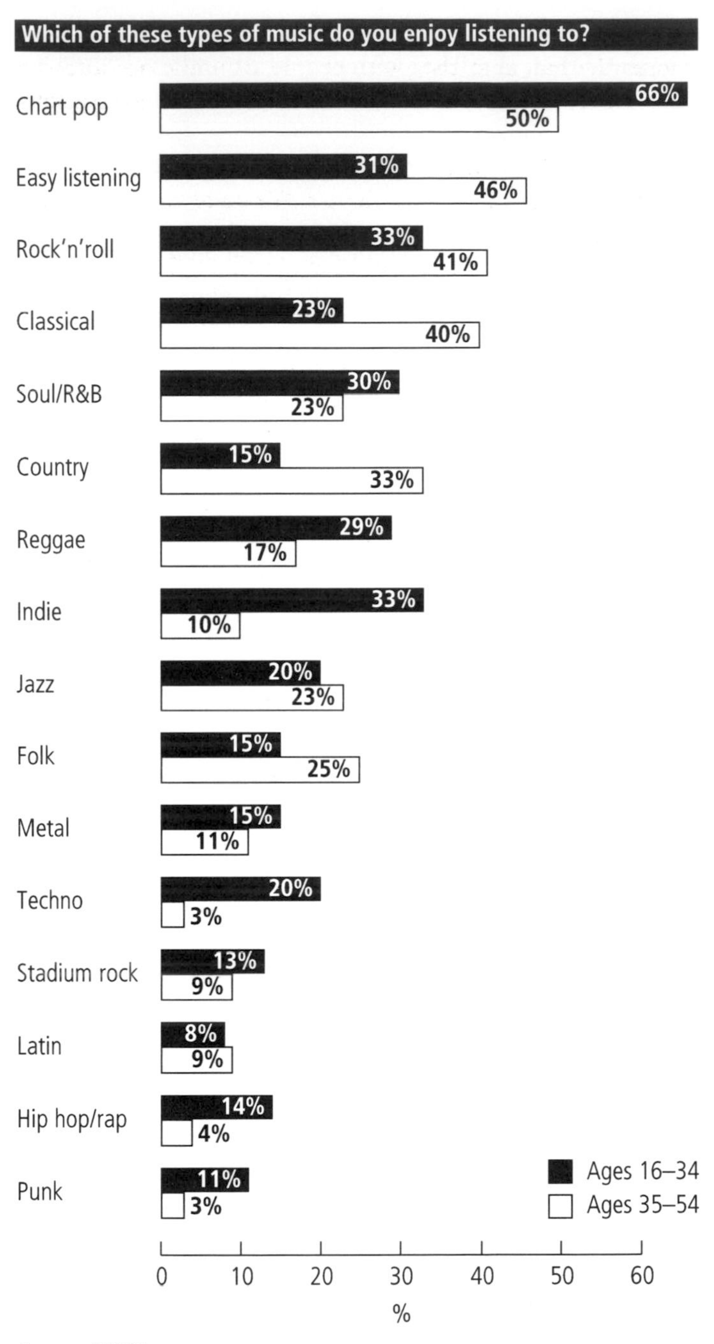

Source: MORI.

Fig. 9.1 Audience appreciation: enjoyment of music by categories and by age

Figure 9.1 shows another way of looking at audiences and effects. You could also look at ways in which marketing strategies relate to the information in this chart – how are these different audiences targeted?

audience does not rationalize about the effects of messages about the value of ageing creams. It ignores the fact that the product cannot work because the skin cannot rejuvenate cells. The effect is to create anxiety and aspirations about appearance.

Agenda Setting: the media have the effect of setting up an agenda of important topics through news activities in particular. We come to believe that this is what the agenda ought to be.

Socialization: the media have the effect of socializing us into the norms and values and accepted behaviours of our society (see also cultural effects and media functions).

Social Control: the media have the effect of controlling the audience by advancing arguments for consensus and for law and order, and by suppressing arguments and materials which question the ways in which our society operates.

Defining Reality: the media define social reality for us. Social reality is what we take to be real, normal and proper, concerning the way that we run our society, and the way that we conduct our social relationships with other people.

Endorsement of Dominant Ideology: the media do have the effect of endorsing this dominant way of looking at the world, this dominant view of power relationships between groups in society, this dominant view of how things are run. To this extent the media become an important part of hegemony – the social and cultural power which the elite and privileged members of society have over the rest. Hegemony naturalizes this power to the extent that it becomes invisible. No one stops to question notions of high art or of good taste, as taught in schools or colleges. The power of hegemony creates assumptions – for example, perhaps, that Channel 4 has a 'better' type of programme. Similarly, the class structure of the social system tends to be taken for granted. Ultimately, the concept of hegemony underpins Parliament itself, creating an assumption that what it does must be right and good for society as a whole.

This account of effects is generally critical of the media and of institutions associated with it. As a view of influences it is broadly in line with the views of those social theorists known as the Frankfurt School, who certainly saw the media as an agency of social control and as undermining the capacity of the audience to think critically about the media, or indeed about any other insitution. This position is much the same as when one looks at dysfunctional effects of media systems. It should be said that there are opposing and positive views of the constructive functional influence of the media. In this case it would be argued for example that the media have the effect of maintaining cultural individuality and of bringing it to a wide audience. This in turn could be opposed to the view the the media undermine and deny cultural and subcultural differences by creating a mass culture.

However, at this point I need to remind you that both views tend to be underpinned by an assumption that the media do things to helpless audiences. The last chapter suggested that people are not always that passive or unaware of what may be going on.

Finally one may usefully look at effects as categorized by McQuail, in respect of beliefs and opinions held by the audience (*Mass Communication Theory*, 1994):

- cause intended change (conversion of views)
- cause unintended change
- cause minor change (form or intensity of views)
- facilitate change (intended or not)
- reinforce what exists (no change)
- prevent change.

2 STUDYING EFFECTS

2.1 Methods of Research

First of all it is worth looking briefly at some ways of investigating and measuring effects on the audience. This presumes we know what we want to measure. Usually this has to do with behavioural change, or most likely nowadays, attitudinal change. In other words, **can we prove that the media** (or some part of their output) **have affected the way that people act or think?** How can we prove this?

There is an attraction in quantitative methods of research. These ask very specific questions which demand closed responses, the simplest of which is yes or no. They appear to be factual and objective in their approach. Their attraction lies partly in being able to achieve a statistical response, in being able to talk about percentages. However, qualitative research is just as valid, even if it deals in looser and more immeasurable factors, such as how people feel about a particular media product.

Research can deal with the audience by questionnaires and interviews. It can deal with the product/material separately by straight analysis. It can deal with the interaction between the media and their audience, **either first hand or through replication of the viewing/reading experience**. The Nielsen market research organization in the United States supervises film previews during which the sweat and pulse rate of the audience can be measured, who also turn knobs by the viewing seat to register their like or dislike for certain scenes or characters. Replication asks sample audiences their opinions of material shown to them. This is of course not the same as seeing the communication on the screen or at the point of sale. The difficult thing is to make the experience authentic and therefore the reactions truthful and therefore the evidence accurate. There are various other details of method which attempt to make the evidence objective. The classic detail is the use of a **control group**, which seeks opinions from those who have not heard the radio programme in question, and then compares their reactions with those of a group which has.

Closed Experiments are those where the research may be done in a laboratory or a very specific place and is concentrated on particular aspects of the material and of audience reaction. For example, children have been shown violent acts (sic) on screen, still or live, and have then been monitored for their style of play behaviour. The experiment is controlled, but it is also essentially artificial because normally we read and view in many places with all sorts of things going on around us.

Field Studies are those where researchers go out to the audience, trying to obtain evidence of behaviour as it happens, or trying to obtain opinions and reactions soon after the media experience. One recent example of the former was based on video cameras fitted in TV sets to watch the viewers' behaviour, and then on discussions with these people afterwards. In the case of the latter, classic audience questionnaires would be a staple approach.

Content Analysis has already been described as a quantitative analysis of media material fastening on content (and possibly treatment and structure). In this case one is making particular assumptions about effects on the audience. It is jumping to conclusions to say that because the material contains, say, a high proportion of images of a comfortable material lifestyle therefore it generates a longing for that lifestyle in the reader/viewer.

2.2 Media Audience Research

Our opinions about media material (and issues of the day) are being constantly sought by a number of research organizations both independent and attached to the various media.

The Broadcasters' Audience Research Board (BARB) is jointly owned by the BBC and ITV. It samples which age groups are watching which programmes and when. The audience is sampled across age, class, geographical spread and so on. Three and a half thousand homes are electronically monitored every day to see which channel the TV set is switched to. Three thousand different people are interviewed every day around the country, shifting, so that nearly a million people are sampled annually. Some of the questionnaires are very long, and look into what people like watching and listening to and why. There have been special surveys of audience attitudes to violence on television or their attitudes towards particular channels. But generally BARB and those working in the broadcasting industries are concerned with ratings. It is true that now they are well aware of the importance of numbers within a target audience. The high proportion of young male viewers for American football is important to those selling the advertising space: consumer products such as jeans will be tied to such a programme. It is also the case that the BARB researchers produce what they call an audience appreciation index (AI). But the fact is that, particularly in commercial television, the programming executives don't care much about a high AI index for a programme that is losing its total audience figures. That programme is still likely to be moved out of prime time, or even axed.

The National Readership Survey is funded by newspaper owners and advertising agencies. It conducts 28 000 half-hour interviews a year in people's homes. These cover 200 titles of newspapers and magazines, and deal with reading habits, topic preferences, and audience background.

The Audit Bureau of Circulation (ABC) monitors, via retailers and audience research, the sales and circulation figures for newspapers. An example of an independent research organization would be that of Gallup polls. This kind of 'official' research is clearly exhaustive in its scope, and careful in its identification of audience samples. Yet it is often as much concerned with audience size and with what attracts audiences as it is with audience thinking and attitude formation. Necessarily, it is not seeking views and reactions which could show serious criticism of the material or indeed of the whole structure of media communication as it is set up in this country.

2.3 Problems Surrounding Effects Research

The basic problem with identifying and measuring media effects is that **there are so many variables in the process through which material is constructed, transmitted, and decoded**. There is also the issue of trying to deal with all the media at once – which most research does not. Obviously one is not comparing like with like: different means of communication have different qualities. But then it is also a problem if one ignores all media but one because, as has been said, many messages actually do come through more than one medium at a time. There is a tacit assumption that visual aspects of media are 'powerful' because the images are directly decoded (they are iconic). But one cannot really say that, for example, reading about pornographic acts (sic) in a book does not endorse them just as film viewing is supposed to.

In any case, **the context of communication matters a great deal**. Television is seen as 'powerful' (i.e. influential) because it reaches into the home. But then there may be counteracting influences in the home, where other people can offer opinions on the material. Perhaps reading comics on the bus coming home from school is influential because it is an isolated activity. Then again, perhaps cinema viewing as an isolated activity is not so influential because everyone 'knows' that one goes in to this dark room to experience a fiction, which is not a general part of everyday experience. Apparently context matters, but just how is not clear.

Another problem has to do with the possible **assumptions made** by researchers. I have commented on this with relation to 'official' research. But then academic research is equally open to the error of assumptions. People are likely to approach research with some propositions or hypotheses that they would like to prove. The industries of academe are not perfectly neutral by contrast with the commercial industries of audience research.

There are so many factors which operate on the understanding and judgement of any member of the audience that trying to separate the influence of the media in particular is extremely difficult. People are, for example, influenced by the opinions of family and friends. Certainly, attempts to demonstrate direct or immediate effects are fairly useless. More than 10 years ago McQuail observed that 'numerous investigations have by now seemed to establish that the direct effects of mass communication on attitudes and behaviour are either non-existent, very small, or beyond measurement by current techniques' (*Communication*, 1984: Longman).

Closed experiments tend to be restrictive. They exclude the range of social experience, the complexity of real life as it is lived, and as it must affect the decoding of media material.

Field studies are only as good as their methodology and the interpretation of results. Belson, for example, has questioned children, asking them to remember programmes they have seen some years before (when trying to demonstrate the influence of television with regard to violence). Anyone who has asked people questions about what they watched last night will see the flaw in this one. Barker, during a particular furore about children watching video nasties, demonstrated that children would actually lie about what they had seen in order either to please the researcher or perhaps to boost their own sense of status (*Video Nasties*, 1984: Pluto Press).

Content analysis, as has been said, is essentially divorced from the audience. Frequent use has been made of this approach over the years. But to enumerate the number of shootings on television over a month, or the total acts of physical violence in children's programmes, proves nothing about how the audience perceives these acts.

Research methodology as such needs to be looked at carefully. There was a well-known and still sometimes reported piece of research in the United States into television and social behaviour. This was interpreted especially in terms of violent behaviour and was produced as The Surgeon General's Report. On the face of it a research programme costing one million dollars should have been significant. But other commentators have remarked on the fact that the 23 separate projects involved were not in fact co-ordinated in any coherent way, that there was no distinction made between social and personal violence or collective and individual violence, that there was no sense of the social context of the violence described.

So there are clearly considerable problems surrounding the administering and interpretation of research into effects.

2.4 Conditioning Factors Around Effects

Numerous factors affect how messages are understood from media material, and how the material affects the audience. Important ones are as follows:

The Source

This concerns the nature of the power held by those who deliver the messages. This refers especially to those who create the material and who present it to the audience (as opposed to the owners who control a given medium). For example, there is **referent power** when viewers may identify with a presenter as someone of their own 'type' and with their views. They will then be predisposed to believe what that person says. Or there could be **expert power,** when magazine readers might be inclined to believe what they are told by someone either billed as an expert or whom they believe (probably from reading about them!) to be an expert.

Similarly, there is the concept of **opinion makers** in the media. These are respected individuals who are seen to be knowledgeable and trustworthy, and so are readily believed. Newsreaders or current affairs presenters would be examples of such people. The whole style of their presentation, the history of their appearances in the broadcasting media inclines the audience to accept what they say, not to question it.

The medium as source also makes a difference. Radio news programmes are perceived as being more 'reliable' than, say, a satirical programme based on news items. The judgements and criticisms made by the latter could be piercing and worthy of consideration. But the former will be seen as having more status.

The monopoly enjoyed by the source will make a difference. For example, a great deal of hard news material especially in regional papers comes from the Press Association (collectively owned by the major regional newspaper groups). Regional papers cannot afford to have their own reporters in the field. They buy in their news. Regional populations receive essentially the same hard news from this source. There is a process of repetition and reinforcement going on. The audience has, because of this degree of monopoly, a relatively limited ability to check out what they are told, to seek alternative versions of main news. So the effect of presenting a particular view of the world is strengthened.

In the end, the crucial factor is the credibility of the source.

The Context

There is the concept of **opinion leaders** who fit into a social context. Research has demonstrated that **most people within their social groups are inclined to listen to certain respected individuals** (it could be your mother or your best friend or your teacher). If these people endorse some opinion articulated through the media, or offer some opinion on the material, then the audience is inclined to listen to them. If these people confirm a media opinion or endorse the value of some programme or magazine, then the audience will be the more affected by that material.

Social conditions will affect how far some media item will influence the views and values of the audience. If the audience member has been a victim of some crime in an inner city area which is plainly suffering from neglect, then the likelihood is that they will pay attention to media information and views on the subjects of urban crime and urban decay.

Media events and media issues are also part of the context. This is a kind of self-fulfilling prophecy, where the media condition their own effects by setting up conditions. For example, in 1989 we had a spate of stories about problems with food hygiene. The sheer quantity of material put this issue on the agenda, made it an issue in the first place, and also predisposed people to pay more attention to further items on the subject. An atmosphere of anxiety, and criticism of the monitoring of food hygiene, predisposes people to feel anxious and critical when any new information on the subject comes their way.

The Message

The first point to be made here is that **what is not said, what is not known by the audience because it is not said, will condition the effects of the**

communication. For example, one might buy a record by a favourite band or orchestra and listen to it as one had done before. What you don't know is that this group of musicians is going to break up and will never record together again. If you knew that, then your attention to the music and its effects would be different.

Repetition of messages tends to enhance their effects. People tend to believe something if it is said often enough (providing it isn't too outrageous!).

Another conditioning factor is **whether or not the message is about something which has already been presented in the media.** In other words, once something has been made an issue – such as aircraft crashes and air safety – then another item about this subject is likely to get attention because it fits a pattern. Again, there is a kind of irony that the media themselves have created this heightened awareness.

Then there is the matter of **the use of conventions in presenting the message.** It is obvious that if a news item is presented through banner headlines and dramatic photographs then this will condition the readers' responses. Similarly, the conventions of visual treatment in comics incline the audience to lock into a fantasy mode in terms of realism, and to expect humour or satire.

There is the question of **the message's structure**. This could refer to the way a story is told in a magazine or to the composition of a photograph or to the structure of a piece of music. To take a simple example, if one has a photograph in a brochure which carries a message about happy holidays, and the children playing in a pool are out of focus (in the background), the effect of the message will be different than if they are clearly in the foreground. There is a shift of emphasis.

■ *The Audience*

There are a number of conditioning factors here. The **place of the audience in the social structure makes a difference.** Those who are well off are likely to be affected by a message about increases in taxation differently from those who are badly off.

An important factor has to do with **what the audience believes already.** Someone believing in self-sufficiency and independence will respond to an advertisement encouraging people to start up small businesses in a different way from someone who believes in job protection and a job for life. **The greater the match between the views and knowledge of the audience and what is said in the communication, the more the communication will be believed.**

The needs of the audience at one time or over a period of time will condition how it is affected by given items. If one is feeling depressed and lonely then one will interpret a situation comedy about the trials of life on one's own in an apartment differently from someone who can't wait to get into their own place.

It must be true that the way the audience perceives will condition effects. **Ultimately it is all about perception.** All media material like all communicative experience has to be perceived. The experience, the material, the person, has to be judged and assimilated into one's way of looking at things. To pick up the last example, one person will view a sitcom scene showing a man making a mess of domestic chores in one way – perhaps perceiving it as not too serious a comment and typical of male incompetence anyway. Another audience member might perceive it as an annoyingly stereotypical view.

Morley (1992) discusses his work and that of others when he takes a view of how audiences may decode or read texts. The notion of a **dominant reading** is much the same as that of a preferred reading, where he argues that certain social groups will make sense of a text in the way that it 'wants' us to make sense of it, a way that fits the dominant ideology. Those who make a **negotiated reading** are able to stand back from the text to make certain choices in terms of what it may be saying. **Oppositional readings** are those which actually make sense of the text in quite the opposite way to how it was intended to be taken. The research suggests, not surprisingly, that what we make of a text depends on where we are coming from. For a start, you may develop a capacity to negotiate readings with texts as a result of reading this book and taking on board at least some of its implications. But in any case you will already have certain positions, sets of attitudes and values, in your head as a result of your upbringing and of the norms of those groups which you value. To pick up an example from Morley, you are going to look at a television news piece about some industrial dispute rather differently if you are a trainee manager in a bank, than if you are a trade union official working in a factory.

3 REASONS FOR STUDYING EFFECTS: EXAMPLES

It is worth noting that **when effects are discussed there are often specific issues raised**, such as the effect of television on children, or the effect of television on family life. Whilst it is useful to fasten on a particular issue and a particular audience, there are a lot of assumptions behind all this.

In both cases the exclusion of other media ignores important influences. What about comics or videos where children are concerned? What about the effect of tapes and records on family life?

3.1 Children and Television

Propositions about children in relation to television need to take account of many things. What do we mean by a child? The programmers or the film censors clearly realize that **different age ranges are likely to be affected in different ways**. Again, there is a weight of evidence which shows that children are able to distinguish between modes of realism by about the age of eight. So cartoon violence or indeed any fictional violence is less likely to affect the older child because it is seen as being of a special world apart from the one in which they live.

There is also **concern about various kinds of role model offered to children through television,** not least in programmes made for children. *Grange Hill* is a well-known British series about a fictional state school. There have been anguished comments about its depiction of elements such as bullying, young

TVRs	All children
Honey, I Shrunk the Kids	41
Gladiators	38
National Lottery	37
Do it Yourself, Mr Bean	36
Neighbours	36
	Girls
Neighbours	40
Honey, I Shrunk the Kids	39
3 Men and a Little Lady	38
EastEnders	37
Gladiators	37
	Boys
Honey, I Shrunk the Kids	42
National Lottery	41
Ghostbusters II	39
Gladiators	39
Do it Yourself, Mr Bean	37
	4–9s
Honey, I Shrunk the Kids	46
Gladiators	40
ET	40
Ghostbusters II	37
Do it Yourself, Mr Bean	36
	10–15s
Neighbours	42
National Lottery	38
EastEnders	38
Twins	37
Casualty	37

Source: BARB/AGB in *Spectrum*, Spring 1995, ITC.

Fig. 9.2 Children's viewing – most popular programmes, 1994

Figure 9.2 shows the top TVR ratings for programmes which children like to watch. You could draw conclusions about their tastes and interests. You might also consider the ideology of the programmes concerned, and the potential effects on these young people.

romance and racialism. One proposition is that children may adopt the worst role behaviour and the least attractive attitudes from the drama (even though it takes a clear moral stance). This is about effects as imitation and effects as attitude change. There is no evidence that either effect or consequence takes place. It does seem that adults are uncomfortable with a relatively truthful view of the child's school world which does not fit their ideals. Children have been heard to comment that reality can be worse, that the programme makers have cut out the swearing anyway!

What is said about children and television viewing (and its effects) often says more about the adult commentators than it does about the children. This is not to argue that parents should not have attitudes and values regarding the content and quality of television, that these parents should not try to impart these values to their children. One would expect this to happen as part of the usual process of parental socialization. But these parents underestimate their children if they believe that they have no powers of discrimination or judgement. They are overestimating television and its effects if they really believe that, on its own, it is going to make their children passive or in some way 'control their minds'.

3.2 The Family and Television

Propositions with regard to television's effect on the family include assumptions which see the family as divided by programme preferences and struck dumb by their engagement with the screen. This ignores a reasonable and opposite assumption – that the medium gives the family something in common to talk about. Investigations show that this does happen in at least some families. They show that **families certainly do not sit quietly in front of the screen, each member in a private world. People talk about what is going on.** Women in particular, it has been found, are more lively and active than men in discussing what they see and what it means, as well as being able simply to ignore it.

When people argue from their own experience that television is divisive it needs to be considered that perhaps television is the symptom of the division, not the cause. The husband immured with weekend sport may be using it to express alienation from his partner: there is no need to assume that television has caused this situation. In any case, there is also evidence that members of the family simply do not fall slaves to television. Young people in particular may watch their cult programmes, and talk about them animatedly, but they also leave the box behind and go out and do other things.

3.3 Violence and Television

Many people have opinions about this subject, but little has been proved. In one survey the IBA found that 60 per cent of sample viewers asserted that there was too much violence on television. But when questioned more closely very few of these people could actually name specific examples. Another IBA survey asking about a possible decline in the standards of television found only 5 per cent who thought this was because of excessive violence. Yet, another survey by the IBA (December 1987) suggested that as many as 6 per cent of viewers sometimes felt violent after watching violent programmes on television. The actual question was about whether or not the viewer agreed with the statement 'sometimes I feel quite violent after watching crime programmes'. Apart from the fact that violence is not confined to crime programmes anyway, the answers to such questions in no way tell us how long the feeling lasted, nor could they prove that these people then behaved violently.

The media themselves are prone to raise media violence as an issue (often asserting that it is excessive), and so perhaps falsely affect public opinion of how much violence there really is. It was notorious that the popular press asserted that the killer behind the 'Hungerford Massacre' of 1988 was influenced by violence in Rambo films. There was no evidence that the person concerned had ever seen these films. Still this notion of effect through 'imitation' remains a popular one.

These comments are not meant to ignore people's anxieties. There is concern about the long-term effects of the media – perhaps that of desensitization, especially where children are concerned. But in terms of there being a connection between television portrayals and human behaviour, then as the Director General of the IBA said in 1987, 'a direct causal link has never been convincingly established'.

As I have said earlier in this chapter on possible media effects, there are many conditions and influences which surround possible effects. A survey by Portsmouth Polytechnic lecturers reported in *New Society* (18 August 1987) suggested that, for example, where people live relates to violent behaviour as much as what they view.

I have also pointed out that research methods (and even researchers' possible predispositions) can cause problems. Belson in *Television and the Adolescent Boy* (1978: Saxon House) purported to find a connection between violent behaviour and the viewing of violent programmes when people are young. But the problem was that he was looking at youngsters chosen because they had committed crimes in the first place, and then asked them for comment based on a list of programmes which seem to have been pre-selected because many of them were relatively violent compared with much television material. Such selection of samples and of materials was bound to produce a correlation.

One survey of violence on television – *The Portrayal of Violence in BBC Television* (Cumberbatch, 1989: BBC) – is very useful in its content analysis approach, and to this extent is objective. It appears to prove some interesting things; for example that, measured in terms of acts of violence, British TV is actually growing less violent, and is certainly less violent than US television. What the study does not and cannot do is to connect the evidence covering types of violence and their performance in certain types of programme with actual violent behaviour by the audience.

Another point you might consider with relation to effects on children in particular is that concerning the distinction between fact and fantasy. Children may talk violently or play violently, to an extent, after watching certain television material. But this does not mean that they are going to think violently or act violently some time after this experience is over.

In 1985 some research into children and their viewing of violence suggested that as many as 40 per cent watched so-called violent programmes (films) sometimes. In 1988 another team replicated this research, but in their questioning slipped in titles of films which did not exist. 68 per cent of children surveyed claimed to have seen these non-existent films. So again one has doubts about the supposed links between types of violence in the media and violent behaviour in life.

Those who choose to believe that television shows too much violence will continue to believe this. One may remain concerned about the possible effects of what violence there is on television. But you as a student of the media should be cautious about accepting the assertions of researchers without looking at their methods. You should ask yourself basic questions about how one defines violence in the media (what about psychological violence, for example?), about how one measures the extent of such violence, about how one measures audience reactions to that violence and about whether any of this proves a connection between the screen and the mind of the viewer.

3.4 Attitudes, Beliefs and Values

The fact that one can question the validity of research into the effects of the media does not mean that they do not have any effect.

What we have to take account of are two things, in the main. One is that the **communication between the media and the audience is extremely complex**. Propositions about and discussions of effects have to take account of a range of factors which will influence the process of the production and interpretation of meaning. The other major point is that **one must take account of indirect and long-term effects** through which the media may affect the way that people think about themselves and their society. Equally, the audience will have some reciprocal effect on the media producers. The success of a film such as *Silence of the Lambs*, which is about a serial killer, certainly told producers that this was a subject that the audience would ‘accept’. And so followed successful films such as *Seven*.

In the end, one must return to the matter of **beliefs, attitudes and values. It is their meanings which are at the heart of the communication process. Media Studies is not simply about facts** – who owns what, how newspapers are made, how censorship operates. **It is about the significance of those facts: how we think and how we live out our lives. It is about the values by which we live, and the beliefs which inform our actions. It is about our attitudes towards one another.** The media play a part in constructing all these things. This is why we should study the media.

REVIEW

You should hove learned the following things from this chapter on media effects.

- it is often proposed that the media affect our attitudes, beliefs and values, directly or indirectly.
- there has been a great deal of research into effects on the audience, but much of it is inconclusive or depends on certain conditions.
- there are many factors which may influence people's attitudes and behaviour. The media are only one of these. It is difficult to separate the media from those other factors when conducting research.

1 WHAT SORTS OF EFFECT?

Types of effect may be summarized as follows: short-term effects, inoculation theory, two-step flow theory, uses and gratifications theory, long-term effects, cultural effects, attitude change, cognitive change, moral panics and collective reactions, emotional responses and personal reactions, agenda setting, socialization, social control, defining reality, endorsement of the dominant ideology.

2 STUDYING EFFECTS

2.1 Methods of research notably include closed experiments, field studies, content analysis.

2.2 The media conduct a great deal of audience research of their own. Examples of organizations doing this are the Broadcasters' Audience Research Board (BARB) and the Audit Bureau of Circulation (ABC).

2.3 Problems surrounding effects research, and trying to prove effects, notably include the following: the number of variables to take account of; the context in which communication takes place; assumptions made by researchers; separating the influence of the media from other factors; the unrealistic nature of closed experiments; the quality of methodology and interpretation of field studies; the separation of content analysis from the audience.

2.4 Conditioning Factors Surrounding Effects on the Audience

The status of the source of the message matters. Media authority figures known as opinion makers also matter. The essential nature of the medium influences how the audience is affected. The degree of monopoly of information or of material in general possessed by the medium (owners) matters. In general the credibility of the source matters.

The context in which the messages are received influences how they may affect the audience. Opinion leaders in social groups may be part of that context. Social conditions matter. Events and issues raised by the media themselves are also part of the context in which the messages come through.

The nature of the message matters. What is not said is as important as what is actually said. Repetition has an effect. Messages about subjects which are already in the media matter. The use of conventions of presentation affect the impact of the message. How the message is structured matters.

The audience's powers of perception, their beliefs, their knowledge, influence how messages affect them.

3 REASONS FOR STUDYING EFFECTS

Some discussion of effects is related to specific issues and audiences – such as the effect of television on children, or on family life.

3.1 The effects of television on children depend on many factors, such as the age at which they distinguish fiction from reality, and on preconceptions which adults have about children.

3.2 The effects of television on the family may be constructive, providing a focus for family life and discussion, as much as destructive.

3.3 Violence in the Media: Particular Effects

- there are many opinions offered about this but no direct relationship between media material and habitual violent behaviour has been proved.
- the media themselves often raise violence as an issue and perhaps cause anxiety about it.
- the most plausible effects proposed are long-term and to do with attitude change and de-sensitization towards the depiction of violence.
- there have been many problems with research methods in terms of proving the effects of violence.
- recent content analysis demonstrates that in terms of acts of violence depicted on television there are less than there used to be and less in Britain than elsewhere.

3.4 Conclusions

- problems with research do not mean that the media do not affect us and that we should not be concerned. In terms of media effects we need to be most concerned with possible effects on our attitudes, values and beliefs

Activity (9)

This activity is concerned with the notion of media effects. It is focused on children as subject matter, not least because cultural concerns and assumptions focus on the young audience more than others. It is also often the case that one of these assumptions is that media must have negative effects on the child as audience.

There is a court case in the USA where a defending attorney is trying to hold the film *Natural Born Killers* responsible for the fact that his two young (but not child) clients went on a killing rampage.

You will be asked to take a more positive approach to possible effects. This activity also embraces media functions, and the notion that we obtain kinds of gratification from media experience.

This activity concentrates on children's reading – of books, comics, magazines, any texts where writing is important even if it is not the only code. You will need to decide what age band(s) of children you want to deal with. But don't try to do too much. It is better to do less but well and in depth, than a lot superficially, so that your results lack validity. The basic task is to –

CONDUCT STRUCTURED INTERVIEWS

First **with parents and teachers**, to deal with the following central questions, elaborated as you wish:

- what do they think children read?
- what do they think children should read?
- why do they think children should read?

Then, **with the children themselves**, working within the chosen age band:

- ask what they do read and why.

Please remember that you must let them speak, and offer no opinion or judgement, other than encouragement to go on talking.

Your questioning should not fall into the trap of assuming that reading is about fiction, nor even just about books for young people. Children who have particular hobbies will read quite difficult factual books, but may give stories a miss.

In your follow-up to these interviews you can:

- pick out the similarities and differences between the adults' views and those of the children;
- notice the differences between what adults think children read and what they actually do read;
- see if any of the reading or suggested reasons for it fit in with ideas about media functions;
- see if any of the children's reasons for reading fit in with the uses and gratifications theory;
- note where reading is connected with other media – e.g. following up a topic first found on television, reading the book of the film that was enjoyed, reading a spin-off from a comic. You can decide whether or not this kind of media-connected reading forms a significant proportion of all reading.

Glossary of Terms

The following is a list of terms used in Media Studies, with a brief definition for each one. Most of them appear in the book, but there are one or two which I have added. You may come across them in your further reading, especially if you take on more 'difficult' books. For further references and information see *Key Concepts in Communication and Cultural Studies*, by T. O'Sullivan, J. Hartley, D. Saunders and J. Fiske.

Agenda Setting refers to the process by which the news media define which topics (the agenda) should be of main interest to the audience, by selecting these topics repeatedly for news material.

Anchorage refers to the element of a picture which helps to make clear (anchors) its meaning. In many cases it is actually the caption of words underneath which does this.

Attitude, attitude formation refers to the hostile or friendly view that we hold towards a person or an issue presented in the media. Attitude formation refers to the ways in which the media shape our views on a given topic.

Audience defines the receivers of mass media communication (also readers and viewers). It may be measured in terms such as numbers, gender and spending habits.

Bias describes a quality of media material which suggests that it leans towards a particular view of a given issue when it should be more neutral. The term is often discussed in relation to news treatment.

Binary Oppositions refers to a frequent pattern in texts where characters, ideas or plot elements are set one against the other; the text is organized around the idea of opposites.

Campaign describes the organized use of advertisements and publicity across the media and over a period of time, in order to promote such things as sales of a product, the image of a company, the views of a pressure group or political party.

Censorship refers to the suppression or rewriting of media material by some person or committee that has power over those who created that material. The reasons for carrying out censorship are often based on religion (Eire) or political beliefs (Peru).

Channel is simply the means of communication. This general term could apply equally to speech or to television.

Code (*see also* Secondary Code) is a system of signs held together by conventions. Primary codes include speech made up from word signs, or visual communication made up from image signs.

Cognition refers to the mental process by which we recognize what is going on around us and then make sense of it in our heads, for example when we decode a magazine for its meaning.

Consensus refers to the middle ground of beliefs and values agreed within a society. It also refers to the compromise position that is taken on issues raised through the media.

Conspiracy Theories refer to the idea that the media represent a plot by someone to influence our beliefs and attitudes.

Constraints are factors such as lack of finance or legal obstacles which limit the power of media producers to do what they want.

Content Analysis is a way of analysing the meaning and significance in media material by breaking it down into units and measuring how many of each type of unit appear. The more one item appears the more likely it is to have some significance.

Conventions are rules which organize signs in a code, and so help bring out meaning. These rules are unwritten but can be worked out if one observes repeated patterns in human behaviour or in media material. So there are conventions which govern how we talk with one another, and there are conventions which govern how a soap opera is put together. We know these rules subconsciously. This is what helps us work out what a conversation or a soap opera really means.

Culture (see also Subcultures) is a collection of beliefs, values and behaviours that are distinctive to a large group of people, and which are expressed through various forms of communication. It is common to see culture in terms of nations, but in fact culture can cross national boundaries (e.g. Jewish culture). Culture is represented through dress, religion and art forms in particular, as well as through language. The media show these things and so also become part of the culture.

Cultural Imperialism describes the way in which selling the media products of one country (USA) to others is rather like creating an empire of ideas which influence those other countries.

Deviancy refers to the idea that certain groups within society (and representatives of these) are branded as deviants by news in the media. This means that they are shown in varying degrees as odd, rebellious or even criminal. It also means that such people do not fit in with what are also defined as social norms – accepted attitudes and behaviours.

Diegesis is the text itself and everything that is part of its content (rather than how it is expressed).

Discourse is a difficult term used to describe how certain kinds of understanding are created and perpetuated within various institutions in society. These discourses are about ideas and meanings which are made apparent through the use of communication. So there are sets of ideas, and ways of communicating them which are special to medicine, or education, or television or even news in particular.

Dysfunction describes negative or destructive ways in which the media can be operated or used.

Feedback is communication in response to a previous message. In terms of mass media, feedback is often delayed because those taking part in the process of communication are not face to face. It is also largely ineffective because the audience giving the feedback has no means of insisting that the producers should change their communication in some way in response to that audience.

Gatekeeping refers to the way in which information may be filtered and selected by key individuals in an organization. With reference to news in particular this term draws attention to the fact that we only receive an approved selection of all the news available.

Genre refers to types of media product which are recognizable through having a number of common and identifiable elements which add up to a kind of formula for the creation of story and characters. Genres are also by nature popular and commercially profitable. Many genres are also narrative fiction (e.g. the detective story), but not always such (e.g. quiz shows).

Hegemony describes the exercise of power by elite groups in a society over the rest of that society. The institutions of the media are important in expressing and maintaining that power. This is especially true because the exertion of that power is not necessarily obvious – e.g. it may not be noticed that the media still present it as natural that people who are important in our society are white, middle-class and speak with received pronunciation.

Icon is a symbolic element within a genre which is highly charged with meanings relevant to that genre. An icon is in effect a symbol for the genre – e.g. the trench coat and trilby hat for the detective genre.

Ideology is a term which refers to the coherent set of beliefs and values which dominate in a culture, and which is particularly held by those who have power. Ideology is concerned with social and power relationships and with the means by which these are made apparent. The media communicate ideology to their audiences. This ideology can be found in the material by looking for covert messages.

Ideological State Apparatus refers to those institutions controlled by the state which act as a power for representing (and communicating) this dominant ideology – e.g. education and schools. It is not necessarily obvious that they are doing this because the values comprising ideology are often shown as being naturally correct and therefore made 'invisible'.

Image analysis is the analysis of the signs within the image (photography, painting, cartoon) in order to deconstruct the meaning which these signs create.

Impartiality is the notion of not taking sides or a particular view on a given issue. Broadcast news in particular believes in the idea of impartiality though it does not always achieve it.

Institution is the word identifying any type of media business or organization which owns the means of making media material.

Intertextuality describes the ways in which texts and their meanings are intertwined, from one medium to another or perhaps within a genre. We understand one text partly from what we know about others.

Mass Communication is the general term for means of communication which operate on a large scale – i.e. in terms of the geographical area reached, the numbers of people reached, the numbers of pieces of communication which are reproduced.

Meaning is what you think it is! The meaning of communication, printed or broadcast, is what you believe it is trying to tell you. The meaning in a piece of communication is signified by the signs which make it up. The meaning which is intended by the sender of the communication may not be the same as the meaning which is decoded or 'read' by the receiver or audience.

Media Power refers to the power of media institutions through ownership and control to shape messages which in turn may influence the attitudes and behaviour of the audience.

Mediation describes the way in which the media come between the audience and the original material on which the programme or printed matter is based. The media provide their own version of this original material. In mediating it they change it.

Message is any item of information, of opinion, of value judgement which we believe we are being told through an example of media material.

Mode of Address refers to the way in which any media text 'talks to' the audience, to how it sets up a relationship with the audience through the way it 'talks'.

Moral Panic is a term that describes a collective response to events in society as shown through the media. The original work that helped coin the term was about news coverage of mugging. The suggestion was that the news people created a panic about mugging by reporting it frequently, and implying that it was a more serious problem than it really was.

Myth is a story or even part of a story which is not really true, but which says something about what a culture wants to believe in. Genres are full of myths about heroic deeds or even about ideas – such as the American dream that everyone can be a great success if they really want to.

Narrative refers to storymaking and story structure. The narrative of a programme or an article is not just its storyline. It is also about how the story is organized and about how the understanding of the reader is organized by the ways in which the story is told.

Naturalization refers to the way in which the messages within much media material are made to seem naturally 'right' or 'true'. The messages which make up ideology are usually held to be naturalized, so that one is not aware that the ideology is there.

News Values are the news topics and ways of treating those topics that the news makers judge as important.

Opinion Leaders are those people in groups that we belong to whose opinions we respect, and which we are likely to agree with because of this. The effects of the media depend partly on these opinion leaders.

Opinion Makers are media figures – such as the presenter who fronts a documentary series – who thus have the power to interpret information and represent views to the audience.

Paradigm is a collection of signs which 'belong together' but which then need conventions to turn them into a working code. The alphabet is a paradigm of letter signs which need conventions to organize them into words (syntagms).

Preferred Reading refers to the way in which an article is so written or a programme is so constructed that the audience is subconsciously pressured into preferring (reading) one meaning into that material rather than any other meaning.

Propaganda refers to communication which is intentionally manipulated by a powerful source in order to project influential political messages. This source is also powerful enough to exclude alternative messages. It is a misuse of the term to call advertising propaganda.

Realism is a quality of a piece of communication, fictional or factual, which causes the reader or viewer to judge it to be more or less truthful or lifelike.

Representation refers to both the creation of a likeness of something through using signs and to the creation of meanings through those signs. Representations of people are constructed through image signs in the media. What those images mean or stand for is also represented.

Role is a particular pattern of behaviour that we take on because we see ourselves in a certain social position in relation to other people. Roles are also taken on because others assume we should do this. The positions are defined in terms of factors such as job or place in the family. So we talk about the role of daughter or the role of reporter.

Secondary Code is a particular code of signs and conventions which works on a level above the primary codes of speech, non-verbal communication, etc. The media are full of such codes, which belong to genres in particular – e.g. the news.

Semiotics refers to the study of signs, of sign systems and of their meanings.

Sign: anything can become a sign if we agree that it has a meaning. The particular meaning of a sign or set of signs is something that is agreed and learned through experience. Signs have no particular meaning of themselves. So gestures or elements of pictures become signs because we agree that they have a meaning.

Signification is what the receiver believes that a sign actually does mean, out of all the possible signifieds.

Signified is that part of the sign that is its possible meaning. In many cases the signifer could refer to a number of possible meanings.

Signifier is that part of the idea of sign that actually signals something – a wink, a camera angle, a word.

Socialization: this describes the processes through which a person becomes a participating member of society. It is learning to live with others. It includes the idea of acquiring beliefs, norms, values, conventions, and of ways of interpreting experience – including our experience of the media. The media help to socialize us.

Source is where the message comes from.

Stereotype is a generalized and simplified social classification of individuals or groups. This classification includes untested assumptions and judgements about types of people. So stereotypes are also often about prejudices or negative attitudes towards the type of person or group represented.

Structural Analysis is a process of looking for the structure or organizing principles within a text. (*See also* Textual Analysis, Binary Oppositions, Narrative.)

Structuralism is an approach to studying experience which looks for organizing principles and structures within that experience. In the case of the media it assumes that there are structures to be found in media products. If one finds those structures it is believed that this helps us to find out how the programme or printed item (the text) comes to have meanings.

Subcultures are cultural groups within a main culture. Often they define themselves by opposing that main culture in various ways. For example, Rastafarians see themselves as disagreeing with at least some of the values of mainstream British culture, and certainly have distinct characteristics of their own.

Symbol is a type of sign which bears no literal relationship to what it refers to. Words are symbolic. Pictures are generally called iconic because they usually resemble their subject. The word 'tiger' doesn't look like a tiger, but a picture of a tiger does. The word symbol is complicated by the fact that it is used in different ways. So it is also possible to talk about a picture of a tiger being symbolic in another way – the Esso tiger becomes a symbol of that oil company as well as a symbol of strength.

Text describes a piece of communication which may then be analysed for its signs and its meanings. A conversation, a photograph, a programme, an advertisement, a news article may all be treated as texts for analysis.

Textual Analysis is the process of analysing a text to see how it is put together, what meanings are in it, and how they are built into the text.

Uses and Gratifications Theory is about how the media may affect the audience and about how the audience may use media texts. It suggests that the audience actively uses the media to gratify certain of its needs.

Values includes the idea of taking an attitude towards an issue, of seeing particular ideas as valuable and important. The idea of value includes the idea that we have made a judgement about the relative importance of whatever it is that we value. Soaps such as *The Cosby Show* often endorse the idea of family.

Select Reading and Resource List

There are many, many books on the media. But a lot of these are about specialized aspects of the media, and few take an overall look. There are quite a few on areas such as news and effects, but again, many of them are fairly hard to read. This list is deliberately selective, and tries to make clear why the book is worth reading by having a few lines on each. There are one or two marked with a star *, which I think are OK if you are a GCSE course student. Those marked with + are more suitable as general texts for an A level course. These general texts will cover many of the specialist areas as well. Other books are really further reading for other courses and for those who have a strong interest. If you are studying then your teacher can suggest even more background reading to suit the course that he or she has put together. I have added a section which covers any other books which I have referred to in the text. These will certainly be for A level, even degree students.

There are quite a few trade magazines and one or two journals for teachers which are worth getting hold of if you can. The trade magazines will give you inside information on things like advertising campaigns, on changes of ownership in the media industries and on costs. You will need to look for them in school or college libraries, or in the bigger public libraries.

COMMUNICATION AND MEDIA: GENERAL

* Dimbleby, R. and Burton, G. (1990): *More Than Words: an introduction to communication.* 2nd edition. London: Routledge.

Read Chapter 5 for a potted version of major media topics that may help you get into this book if you have problems. Chapter 1 could be useful if you want to understand more about the basic theory of communication as a process.

+ Fiske, J. (1990): *Introduction to Communication Studies*. London: Routledge.

This is an introduction for the able A level student and for teachers. It adopts a semiotics-based approach and has a lot of theory in it. It is very good in what it does, but is not necessarily the first book to read on a communication course.

* Gration, G., Reilly, J. and Titford, J. (1988): *Communication and Media Studies: an introductory coursebook*. London: Macmillan.

Read Chapters 7–11. These are readable and compact, dealing with topics such as persuasion and control in the media, as well as having two brief chapters on broadcasting and the press.

MEDIA STUDIES: GENERAL

Branston, G. and Stafford, R. (1996): *The Media Students' Book*. London: Routledge.

A large, wide-ranging book which covers a variety of key topics and includes activities.

* Dutton, B. and Mundy, J. (1989): *Media Studies: an introduction*. London: Longman.
An attractive book which packs in a lot of basic information at the same time as being profusely illustrated. There are some useful little exercises. It covers important topics, including the language of the media.

Lusted, D. (ed.) (1991): *The Media Studies Book*. London: Routledge.
Although this book is written for teachers, it is perfectly readable by A level students. It covers key topics such as narrative, institution, audience, representation. Obviously it is angled towards helping teachers deal with these topics, but it provides explanation and information along the way.

Masterman, L. (1985): *Teaching the Media*. London: Comedia.
Although this book is written for teachers, students can get a lot out of it. In particular, Chapters 4–7 on determinants, rhetoric, ideology and audience are excellent stuff in a work which is pretty definitive.

+ O'Sullivan, T., Dutton, B. and Rayner, P. (1994): *Studying the Media*. London: Arnold.
A thorough, useful text, which covers most major topics and provides useful examples.

+ Price, S. (1993): *Media Studies*. London: Pitman.
Covers a wide range of topics and concepts for A level courses. Takes some unravelling in places,

Skeggs, B. and Mundy, J. (1992): *The Media*. Walton on Thames: Nelson.
A sociology textbook based on readings or extracts from various sources, and covering key areas of media studies. A very useful little book, not least for the lucid introductions to sections.

Tolson, A. (1996): *Mediations*. London: Arnold.
Offers a comprehensive introduction to media text analysis, covering a wide range of media.

Tunstall, J. (1983): *The Media in Britain*. London: Constable.
This is a fairly mighty work which is comprehensive. It provides some description of aspects of the media and how they work, as well as interpretation of what it describes. There are sections on the different media. There are also useful chapters on bias, power and control.

Media World (weekly)

Baker, P. and Clarke, M. (1989): *Talking Pictures: an introduction to media studies*. Cheltenham: Stanley Thornes.
This is a filmstrip with cassette commentary which takes the viewer through image reading, genre and news.

Hartley, J., Goulden, H. and O'Sullivan, T. (1987): *Making Sense of the Media*. London: Comedia.
This is a pack of booklets on various media topics such as institutions and audiences. It is quite expensive, so it has to be a library copy.

Kruger, S. and Wall, I.: *The Mediafile*. London: Mary Glasgow Publications.
A biannual publication of photocopiable materials for learning about the media.

Team Video Productions: *Making News* – a video tape about the process and content of newsmaking.

—*Action Video Packs: 1 Controversy; 2 Presenters, Soaps and Sport; 3 Viewing, Appearing and Participating on TV* – the titles suggest the content of the three tapes plus teachers' notes. The material has been taken from an educational TV series.

MEDIA OWNERSHIP (INCLUDING DEVELOPMENT)

Curran, J. and Seaton, J. (1991): *Power without Responsibility: the press and broadcasting in Britain*. London: Routledge.

This book is good because it provides a concise and informed history of the press and broadcasting which also brings one up to date with current developments. The chapter on the sociology of the mass media is similarly concise in packing in issues and concepts.

Tunstall, J. and Palmer, M. (1991): *Media Moguls*. London: Routledge.

This book looks at the empires and careers of media owners across Europe. It also looks at the television and news industries in Europe. Some of it is dry stuff, and it will date. But it is still useful and quite unusual for taking a comparative look at media industries.

MEDIA THEORY (INCLUDING EFFECTS AND VIOLENCE)

McQuail, D. (1994): *Mass Communication Theory*. London: Sage.

Although some may object that this is written by a sociologist (as opposed to a cultural studies person, for example), it really is very comprehensive. A number of the topics which I have discussed are dealt with at greater length and with more explicit theory behind them. It must be about the most comprehensive work on media theory.

+ O'Sullivan, T., Hartley, J., Saunders, D., Montgomery, M., and Fiske, J. (1994): *Key Concepts in Communication and Cultural Studies*. London: Routledge.

This is the best reference work for defining terms and cross-referencing them.

Violence in the Media (1988). London: BBC Publications.

A very useful A4 booklet on this subject, published by the BBC and covering a range of relevant topics chapter by chapter.

Cumberbatch, G. (1989): *The Portrayal of Violence in BBC Television*. London: BBC Publications.

This slim book represents research carried out on the basis of content analysis. It includes tables of findings from the research. It leads to interesting conclusions such as the apparent decline of violence on television at a time when public debate inclined to believe there was more.

Cumberbatch, G. and Howitt, D. (1989): *A Measure of Certainty: effects of the mass media*. London: John Libbey.

This small but expensive book is useful for its summary of effects research, including that relating to violence.

THE PRESS

* Butler, N. (1989): *Newspapers*, from the series *Introducing Media Studies*. London: Hodder & Stoughton.

A compact, useful booklet containing basic information and activities.

* Hubbard, M. (1989): *Popular Magazines*, from the series *Introducing Media Studies*. London: Hodder & Stoughton.

As with the Butler book on newspapers, compact and full of good activities.

UK Press Gazette (weekly)

Willings Press Guide – has addresses and other information relating to these journals.

Benn's Press Directory – gives details of who owns what in publishing, and of newspaper and magazine circulation.

FILM

Giannetti, L. (1993): *Understanding Movies*. Englewood Cliffs, New Jersey: Prentice Hall.
A thorough well-illustrated book – rather expensive – covering all aspects of film criticism.
Nelmes, J. (ed.) (1996): *An Introduction to Film Studies*. London: Routledge.
Useful coverage of major topics, having particular reference to the A level film syllabus.

TELEVISION: GENERAL

Kilborn, R. (1992): *Television Soaps*. London: Batsford.
Although this is actually about a specific TV genre, it is a popular topic. Clearly written, with a combination of information and critical approaches.
* Root, J. (1986): *Open the Box*. London: Comedia.
This book accompanied a television series, some material from which is incorporated in packages available from Team Video. It is compact and generously illustrated. It is interesting because it does not deal directly with traditional media studies topic headings, though it does in the end tell you quite a lot about them. It does deal with items such as presenters, and with the assumptions and prejudices that people have about television.
Selby, K. and Cowdery, R. (1995): *How To Study Television*. London. Macmillan.
This book is not informational or critical in the conventional sense. But it does deal with some key ideas, and it does demonstrate how to deconstruct the texts of television.
Broadcast (weekly)
The Listener (weekly)

MAKING TELEVISION (INCLUDING DOCUMENTARY AND REALISM)

Hart, A. (1989): *Making the Real World: a study of a television series*. Cambridge: Cambridge University Press.
This book also appears as part of a teaching pack (with a video) called *Teaching Television: the Real World*. *The Real World* is a documentary series on aspects of science produced by Television South. The book talks about how the programmes were made, as well as about the knowledge of science that we get from such programmes. It is interesting because it goes further into how our understanding of what science is can be influenced by the way programmes are put together. It deals with narrative and audiences, and with facts about the ownership of TVS and how it comes to produce such a programme series.
Strathclyde University: *Brookside Close Up*.
This is a very useful video tape in two parts which looks at the making of the Liverpool soap and at issues surrounding it, such as the question of its authenticity. Audio Visual Services, University of Strathclyde, Alexander Turnbull Building, 155 George Street, Glasgow G1 1RD.

NEWS

+ Hartley, J. (1982): *Understanding News*. London: Methuen.
This is an excellent and thorough approach to understanding how messages are embedded in news material and how those messages may affect our view of the world. It includes a good semiotics-based chapter on reading the news.

Whitaker, B. (1981): *News Limited: why you can't read all about it*. London: Minority Press Group.
This little book on the press is an interesting read even now, because it tells you all about the ways in which news does or does not get into a newspaper and about all the constraints imposed on a working journalist.

Jones, K. (1985): *Graded Simulations*. 3 vols. Oxford: Basil Blackwell.
The first volume contains Front Page and Radio Covingham, which are classics of their kind, and let you practise working in a newsroom.

ADVERTISING

Dyer, G. (1982): *Advertising as Communication*. London: Routledge.
This compact volume includes chapters on the development of advertising. But it is more useful later on, as the author deals with effects, creation of meaning, semiotics and language in advertising. There are useful examples.

Myers, G. (1994): *Words in Ads*. London: Arnold.
Lively and readable text introduces students to ways of analysing advertisements in their social context.

Vestergaard, T. and Schroeder, K. (1994): *The Language of Advertising*. Oxford: Blackwell.
This is a higher-level kind of book. But it is good at analysis of adverts and their meanings, making connections with important concepts.

Campaign (weekly)

Marketing (weekly)

British Rate and Data (monthly) – directory giving advertising rates for the press and broadcasting, as well as information such as circulation figures.

The Advertising Association: *Finding Out About Advertising*.
A pack of sheets and leaflets which describes types of advertising, gives facts and figures, and includes case studies.

The Advertising Association also publishes sets of leaflets on other aspects of advertising. The Public Affairs Dept., The Advertising Association, Abford House, 15 Wilton Road, London SW1V 1NJ.

The Advertising Standards Authority publishes various booklets, including the Code of Advertising Practice, which covers all print media. The ASA, 15/17 Ridgmount Street, London WC1E 7AW.

Scottish Media in Education: *The Brand X Game* – an advertising activity exercise.

—*Baxters: a study in advertising* – a pack of materials centring on the creation of a Baxters soup advertisement. It includes a video tape. Scottish Film Council, 74 Victoria Crescent Road, Glasgow G12 9JH.

VISUAL COMMUNICATION

Morgan, J. and Welton, P. (1992): *See What I Mean: an introduction to visual communication*. London: Arnold.

Although this book is geared towards visuals (including the media) and is useful in what it says about visual perception, in fact it can act as a general introduction to the subject anyway. It has sections on process and semiotics theory. It covers topics such as myth in genre and uses and gratifications theory.

Bethell, A. (1981): *Eyeopeners 1 & 2*. Cambridge: Cambridge University Press.

These packs contain visual materials designed to help you analyse images so that you can see how meanings are constructed into them by the creator and the audience.

British Film Institute: *Reading Pictures & Selling Pictures*. London: BFI Education Department.

These are two packs of slides, notes and photosheets. They deal with meanings in images and with advertising.

OTHER: VARIOUS

The BBC Annual Report and Handbook – but note that you cannot buy this from the BBC but only from bookshops designated by them.

The ITC Television and Radio Year Book – very entertaining for a trip around their productions.

The ITC Annual Report and Accounts – gives you a lot more hard information about programme hours and expenditures. The commercial section of big public libraries will have this.

The Broadcasters' Audience Research Board publishes reports on TV viewing. BARB, Knighton House, 56 Mortimer Street, London WIN BAN.

The Independent Television Commission publishes various journals and leaflets, most useful of which are *Television: the public's view*, and *Spectrum*. The ITC, 33 Foley Street, London W1P 7LB.

OTHER BOOKS REFERRED TO

Barthes, R. (1977): *Image, Music, Text*. London: Fontana.

Berger, A. (1992): *Popular Culture Genres*. London: Sage.

Brake, M. (1985): *Comparative Youth Culture*. London: Routledge.

Brown, E. (ed.) (1993): *Television and Women's Culture*. London: Sage.

Buckingham, D. (1987): *Public Secrets, EastEnders and its Audience*. London: BFI.

Fiske, J. (1990): *Television Culture*. London: Routledge.

Fiske, J. (1992) in Allen, R. (ed.): *Channels of Discourse Reassembled*. London: Routledge.

Fiske, J. (1994): Television Pleasures. In Graddol, D. and Boyd-Barrett, O. *Media Texts: Authors and Readers*. Clevedon: Open University.

Feuer, J. (1992) in Allen, R. (ed.): *Channels of Discourse Reassembled*. London: Routledge.

Gallagher, M. (1988): Negotiation of control in Media Organisations. In Gurevitch, M., Bennett, T., Curran, J. and Woollacot, J. (eds) *Culture, Society and the Media*. London: Methuen.

Graddol, D. and Boyd-Barrett, O. (1994): *Media Texts: Authors and Readers*. Clevedon: Open University.

Hebdige, D. (1988): *Hiding in the Light*. London: Routledge.

Home Office (1981): *Broadcasting in the United Kingdom*. London: HMSO.

Home Office (1988): *Broadcasting in the Nineties*. London: HMSO.

Hall, S. (1994): Encoding/Decoding. In Graddol, D. and Boyd-Barret, O. *Media Texts: Authors and Readers*. Clevedon: Open University.
Hart, A. (1991): *Understanding the Media*. London, Routledge.
Kuhn, A. (1985) in Cook, P. (ed.): *The Cinema Book*. London: BFI.
Lorimer, R. with Scannell, P. (1994): *Mass Communications, a comparative introduction*. Manchester: Manchester University Press.
McNair, B. (1994): *News and Journalism in the UK*. London: Routledge.
McQuail, D. (1992): *Media Performance*. London: Sage.
Morley, D. (1992): *Television Audiences and Cultural Studies*. London: Routledge.
Tunstall, J. (1993): *Television Producers*. London: Routledge.
Tunstall, J. (1977): *The Media are American*. London: Constable.
Turner, G. (1992): *British Cultural Studies, an introduction*. London: Routledge.